Nonlinear Programming 2

ACADEMIC PRESS RAPID MANUSCRIPT REPRODUCTION

Nonlinear Programming 2

Edited by O.L.Mangasarian
R.R.Meyer
S.M.Robinson

Proceedings of the Special Interest Group on
Mathematical Programming Symposium
conducted by the Computer Sciences Department
at the University of Wisconsin - Madison
April 15 - 17, 1974

Academic Press, Inc.
New York San Francisco London 1975
A SUBSIDIARY OF HARCOURT BRACE JOVANOVICH, PUBLISHERS

ACADEMIC PRESS, INC.
111 Fifth Avenue, New York, New York 10003

United Kingdom Edition published by
ACADEMIC PRESS, INC. (LONDON) LTD.
24/28 Oval Road, London NW1

Library of Congress Cataloging in Publication Data

Symposium on Nonlinear Programming, 2d, Madison,
Wis., 1974.
Nonlinear programming, 2.

Bibliography: p.
Includes index.
1. Nonlinear programming. I. Mangasarian, Olvi L., (date) II. Meyer Robert R. III. Robinson, Stephen M. IV. Association for Computing Machinery. Special Interest Group on Mathematical Programming. V. Wisconsin. University–Madison. Computer Sciences Dept. VI. Title.
T57.8.S9 1974 519.7'6 75-9854
ISBN 0–12–468650–8

PRINTED IN THE UNITED STATES OF AMERICA

CONTENTS

CONTRIBUTORS

Egon Balas (279),
GSIA Carnegie-Mellon University
Pittsburgh, Pennsylvania 15213

Richard H. Bartesl (231),
Dept. of Mathematical Sciences
John Hopkins University
Baltimore, Maryland 21218

Dimitri P. Bertsekas (165),
Coordinated Science Lab
University of Illinois
Urbana, Illinois 61801

Roger Fletcher (121),
Department of Mathematics
The University
Dundee, Scotland

Ubaldo García-Palomares (101),
IVIC
Apartado 1827
Caracas, Venezuela

A. A. Goldstein (215),
Department of Mathematics
University of Washington
Seattle, Washington 98195

Pierre Huard (29),
21 Rés. Elysée I
78170 La Celle-St. Cloud
France

Robert G. Jeroslow (313),
205 GSIA
Carnegie-Mellon University
Pittsburgh, Pennsylvania 15213

Barry Kort (193),
3D-502 Bell Labs
Holmdel, New Jersey 07733

E. Polak (255),
Department of EECS
University of California
Berkeley, California 94720

Michael J. D. Powell (1),
Building 8.9
A.E.R.E., Harwell
Didcot, England

Klaus Ritter (55),
Mathematisches Institut A
Universität Stuttgart
Herdweg 23
7 Stuttgart N, Germany

Kurt Spielberg (333),
38 Knollwood Drive
Cherry Hill, New Jersey 08034

Monique Guignard Spielberg (333),
38 Knollwood Drive
Cherry Hill, New Jersey 08034

I. Teodoru (255),
Department of EECS
University of California
Berkeley, California 94720

PREFACE

In May 1970 the first symposium in Madison on nonlinear programming took place; its proceedings were published the same year by Academic Press. In April 1974, under the sponsorship of the Special Interest Group on Mathematical Programming (SIGMAP) of the Association for Computing Machinery, a second symposium on nonlinear programming was conducted at Madison by the Computer Sciences Department of the University of Wisconsin. These are the proceedings of that second symposium.

This volume contains thirteen papers. Two of the papers (the ones by Fletcher and Ritter) were not presented at the symposium because the authors were unable to be present. In the paper by Powell global and superlinear convergence of a class of algorithms is obtained by imposing changing bounds on the variables of the problem. In the paper by Huard convergence of the well-known reduced gradient method is established under suitable conditions. In the paper by Ritter a superlinearly convergent quasi-Newton method for unconstrained minimization is given. García-Palomares gives a superlinearly convergent algorithm for linearly constrained optimization problems. The next three papers, by Fletcher, Bertsekas and Kort, give exceptionally penetrating presentations of one of the most recent and effective methods for constrained optimization, namely the method of augmented Lagrangians. In the paper by Goldstein a method for handling minimization problems with discontinuous derivatives is given. Bartels discusses the advantages of factorizations of updatings for Jacobian-related matrices in minimization problems. Polak and Teodoru give Newton-like methods for the solution of nonlinear equations and inequalities. The papers by Balas, by Jeroslow, and by Guignard and Spielberg deal with various aspects of integer programming.

It is hoped that these papers communicate the richness and diversity that exist today in the research of some of the leading experts in nonlinear programming.

The editors would like to thank Dr. Michael D. Grigoriadis, SIGMAP Chairman, for his interest and encouragement of the symposium, and Professor L. Fox, editor of the Journal of the Institute of Mathematics and Its Applications, for permission to include the paper by Fletcher. We also would like to thank Mrs. Dale M. Malm, the symposium secretary, for her efficient handling of the symposium arrangements and her expert typing of the proceedings manuscript.

O.L. Mangasarian
R.R. Meyer
S.M. Robinson

CONVERGENCE PROPERTIES OF A CLASS OF MINIMIZATION ALGORITHMS

by

M.J.D. Powell[1)]

ABSTRACT

Many iterative algorithms for minimizing a function $F(\underline{x}) = F(x_1,x_2,\ldots,x_n)$ require first derivatives of $F(\underline{x})$ to be calculated, but they maintain an approximation to the second derivative matrix automatically. In order that the approximation is useful, the change in $\underline{x}$ made by each iteration is subject to a bound that is also revised automatically. Some convergence theorems for a class of minimization algorithms of this type are presented, which apply to methods proposed by Powell (1970) and by Fletcher (1972). This theory has the following three valuable features which are rather uncommon. There is no need for the starting vector $\underline{x}^{(1)}$ to be close to the solution. The function $F(\underline{x})$ need not be convex. Superlinear convergence is proved even though the second derivative approximations may not converge to the true second derivatives at the solution.

1. The class of algorithms

The methods under consideration are iterative. Given a starting vector $\underline{x}^{(1)}$, they generate a

[1)] Computer Science and Systems Division, A.E.R.E. Harwell, Didcot, Berkshire, England

sequence of points $\underline{x}^{(k)}$ (k=1,2,3,...), which is intended to converge to the point at which the objective function $F(\underline{x})$ is least. The components of $\underline{x}$, namely $(x_1, x_2, \ldots, x_n)$, are the variables of the objective function.

At the beginning of each iteration a starting point $\underline{x}^{(k)}$ is available, with a matrix $B^{(k)}$ and a step-bound $\Delta^{(k)}$. $B^{(k)}$ is a square symmetric matrix whose elements are the approximations

$$B_{ij}^{(k)} \approx d^2F(\underline{x}^{(k)})/dx_i\, dx_j \ , \tag{1.1}$$

and $\Delta^{(k)}$ bounds the change in $\underline{x}$ made by the iteration. Both $B^{(k)}$ and $\Delta^{(k)}$ are revised automatically in accordance with some rules given later.

At $\underline{x}^{(k)}$ the first derivative vector

$$\underline{g}^{(k)} = \underline{\nabla}\, F(\underline{x}^{(k)}) \tag{1.2}$$

is calculated. Also we use the notation $\underline{g}(\underline{x})$ and $G(\underline{x})$ to denote the first derivative vector and the second derivative matrix of $F(\underline{x})$ at a general point $\underline{x}$.

To define $\underline{x}^{(k+1)}$ we make use of the quadratic approximation

$$\begin{aligned} F(\underline{x}^{(k)}+\underline{\delta}) &\approx \Phi(\underline{x}^{(k)}+\underline{\delta}) \\ &= F(\underline{x}^{(k)}) + \underline{\delta}^T\underline{g}^{(k)} \\ &\quad + \tfrac{1}{2}\underline{\delta}^T B^{(k)}\underline{\delta} \ . \end{aligned} \tag{1.3}$$

Because most gradient algorithms terminate if $\underline{g}^{(k)}$ is zero, and because this paper studies convergence properties as k increases, we suppose that $\underline{g}^{(k)}$ is never identically zero. Therefore we can calculate a value of $\underline{\delta}$, $\underline{\delta}^{(k)}$ say, such that the inequality

$$(1.4) \qquad \Phi(\underline{x}^{(k)}+\underline{\delta}^{(k)}) < F(\underline{x}^{(k)})$$

is satisfied. Then $\underline{x}^{(k+1)}$ is defined by the equation

$$(1.5) \quad \underline{x}^{(k+1)} = \begin{cases} \underline{x}^{(k)} + \underline{\delta}^{(k)}, & F(\underline{x}^{(k)} + \underline{\delta}^{(k)}) < F(\underline{x}^{(k)}) \\ \underline{x}^{(k)} & , F(\underline{x}^{(k)} + \underline{\delta}^{(k)}) \geq F(\underline{x}^{(k)}), \end{cases}$$

which provides the condition

$$(1.6) \qquad F(\underline{x}^{(k+1)}) \leq F(\underline{x}^{(k)}) \;.$$

In order to obtain fast convergence ultimately, we note that, if the matrix $B^{(k)}$ is positive definite, the least value of expression (1.3) occurs when $\underline{\delta}$ is the vector $-[B^{(k)}]^{-1}\,\underline{g}^{(k)}$. Therefore, if the inequality

$$(1.7) \qquad \| [B^{(k)}]^{-1}\, \underline{g}^{(k)} \| \leq \Delta^{(k)}$$

is satisfied, and if $B^{(k)}$ is positive definite, $\underline{\delta}^{(k)}$ is defined by the Newton formula

$$(1.8) \qquad \underline{\delta}^{(k)} = -[B^{(k)}]^{-1}\underline{g}^{(k)} .$$

Otherwise the length of $\underline{\delta}^{(k)}$ is restricted by $\Delta^{(k)}$, so we force the equation

$$(1.9) \qquad \| \underline{\delta}^{(k)} \| = \Delta^{(k)} \;.$$

On every iteration the choice of $\underline{\delta}^{(k)}$ must satisfy the inequality

$$(1.10) \quad F(\underline{x}^{(k)}) - \Phi(\underline{x}^{(k)} + \underline{\delta}^{(k)}) \geq c_1 \, \|\underline{g}^{(k)}\| \min[\|\underline{\delta}^{(k)}\| , \|\underline{g}^{(k)}\| / \|B^{(k)}\|] ,$$

where c_1 is a positive constant, which is stronger than condition (1.4). It is proved in Section 5 that equation (1.8) satisfies this condition. Here and throughout the paper the vector norms are Euclidean, and the matrix norms are subordinate to the vector norms. Because the constant c_1 in expression (1.10) is arbitrary, this choice of norm does not lose any generality. This remark also applies to the other conditions that are imposed on the class of algorithms.

We have been vague about the definition of $\underline{\delta}^{(k)}$ deliberately, in order to increase the range of applicability of the convergence theorems that are presented. The generality of condition (1.10) is useful because it allows $\underline{\delta}^{(k)}$ to be specified by either the Levenberg/Marquardt (Marquardt, 1963) technique or by the dog-leg technique (Powell, 1970). This statement is proved in Section 5 .

The initial second derivative approximation $B^{(1)}$ may be any symmetric matrix, and any method may be used to define the sequence of matrices $B^{(k)}$ $(k=1,2,3,\ldots)$, provided that the condition

$$(1.11) \qquad \|B^{(k)}\| \leq c_2 + c_3 \sum_{i=1}^{k} \|\underline{\delta}^{(i)}\|$$

is satisfied, where c_2 and c_3 are positive constants. However a stronger condition is needed if one requires the rate of convergence of the sequence $\underline{x}^{(k)}$ $(k=1,2,3,\ldots)$ to be superlinear. A suitable condition is given later.

The general strategy for revising $\Delta^{(k)}$ is that it may sometimes increase, but it is decreased when the actual change in the objective function $[F(\underline{x}^{(k)}) - F(\underline{x}^{(k)} + \underline{\delta}^{(k)})]$ is much worse than the predicted change $[F(\underline{x}^{(k)}) - \Phi(\underline{x}^{(k)}+\underline{\delta}^{(k)})]$. Thus the bound on the change in $\underline{x}$ is related automatically to the accuracy of the approximation (1.3). Specifically we require positive constants c_4, c_5, c_6 and c_7 such that, when the inequality

$$(1.12) \quad F(\underline{x}^{(k)}) - F(\underline{x}^{(k)}+\underline{\delta}^{(k)}) \geq c_4\{F(\underline{x}^{(k)}) - \Phi(\underline{x}^{(k)}+\underline{\delta}^{(k)})\}$$

is obtained, then $\Delta^{(k+1)}$ satisfies the bounds

$$(1.13) \quad \|\underline{\delta}^{(k)}\| \leq \Delta^{(k+1)} \leq c_5 \|\underline{\delta}^{(k)}\| ,$$

and, when inequality (1.12) fails, $\Delta^{(k)}$ satisfies the bounds

$$(1.14) \quad c_6 \|\underline{\delta}^{(k)}\| \leq \Delta^{(k+1)} \leq c_7 \|\underline{\delta}^{(k)}\| .$$

As well as being positive, these constants must obey the conditions $c_4 < 1$, $c_5 \geq 1$ and $c_6 \leq c_7 < 1$. Moreover we require each algorithm of the class to impose a fixed upper bound on the step-lengths, so we have a condition of the form

$$(1.15) \quad \Delta^{(k)} \leq \bar{\Delta} , \; k = 1,2,3,\ldots .$$

The class of algorithms that is analysed consists of the algorithms that meet all the conditions given so far. We claim that these conditions are sensible in practice, and that they allow some useful methods, including an algorithm proposed by Powell (1970) and the version of Fletcher's (1972) hypercube method when there are no constraints on the variables.

In Section 2 it is proved that, under very mild conditions on $F(\underline{x})$, each algorithm of the class provides the limit

$$\text{(1.16)} \qquad \lim\inf \|\underline{g}^{(k)}\| = 0 .$$

Therefore, if one of the points $\underline{x}^{(k)}$ falls into a region where $F(\underline{x})$ is locally convex, if $F(\underline{x})$ has a local minimum in this region, and if the step bounds $\Delta^{(k)}$ and the inequality (1.6) keep the later points of the sequence $\underline{x}^{(k)}$ $(k=1,2,3,\ldots)$ inside the region, then convergence is obtained to the local minimum. Thus it is common for the sequence $\underline{x}^{(k)}$ $(k=1,2,3,\ldots)$ to tend to a limit.

It is proved in Section 3 that, when the points $\underline{x}^{(k)}$ tend to a limit at which the second derivative matrix of $F(\underline{x})$ is positive definite, then the sum $\Sigma\|\underline{\delta}^{(k)}\|$ is convergent. In view of the fact that the conditions (1.11) on $B^{(k)}$ are not very restrictive, this result is quite pleasing, and also it shows that the matrices $B^{(k)}$ are uniformly bounded.

However one would prefer a much stronger theorem on the convergence of the sequence $\underline{x}^{(k)}$ $(k=1,2,\ldots)$, for the algorithms of the class that relate the matrices $B^{(k)}$ $(k=1,2,3,\ldots)$ to the second derivatives of $F(\underline{x})$. For example the updating formula, that is obtained by symmetrizing the Broyden rank-one formula (Powell, 1970), satisfies condition (1.11) and provides the limit

$$\text{(1.17)} \quad \|g(\underline{x}^{(k)}+\underline{\delta}^{(k)}) - \underline{g}^{(k)} - B^{(k)}\underline{\delta}^{(k)}\| / \|\underline{\delta}^{(k)}\| \to 0.$$

Therefore it is proved in Section 4 that the algorithms of the class that satisfy condition (1.17) cause the sequence $\underline{x}^{(k)}$ $(k=1,2,3,\ldots)$ to converge superlinearly. Conversely, the work of Dennis and Moré (1973) shows that when condition (1.17) does not hold, usually superlinear convergence is not obtained.

Section 5 considers the applications of the theory of Sections 2,3 and 4, and includes a proof of the fact that the symmetrized form of the Broyden rank-one formula gives the limit (1.17).

2. A global convergence theorem

It is proved that the following theorem is true, if $F(\underline{x})$ is bounded below, is differentiable, and if $\underline{g}(\underline{x})$ is uniformly continuous.

Theorem 1 The vectors $\underline{g}^{(k)}$ $(k=1,2,3,\ldots)$ are not bounded away from zero.

Proof First we note that, if Σ' denotes the sum over the iterations for which condition (1.12) is satisfied, then the inequality

$$\sum_{i=1}^{k} \|\underline{\delta}^{(i)}\| \leq \left[1 + \frac{c_5}{1-c_7}\right]\left[\|\underline{\delta}^{(1)}\| + \sum_{i=2}^{k}{}' \|\underline{\delta}^{(i)}\|\right] \tag{2.1}$$

holds. This remark is an immediate consequence of the fact that, if inequality (1.12) fails for $i=p+1,p+2,\ldots,q$, then expressions (1.13) and (1.14) imply the bound

$$(2.2)\quad \sum_{i=p}^{q} \|\underline{\delta}^{(i)}\| \le \|\underline{\delta}^{(p)}\| [1 + c_5 + c_5 c_7 + c_5 c_7^2 + \cdots + c_5 c_7^{q-p-1}]$$

$$< \|\underline{\delta}^{(p)}\| [1 + c_5/(1-c_7)] \; .$$

Thus we deduce from inequality (1.11) that there exist constants c_8 and c_9 such that the condition

$$(2.3)\qquad \|B^{(k)}\| \le c_8 + c_9 \sum_{i=1}^{k}{}' \|\underline{\delta}^{(i)}\|$$

is satisfied.

Now the fact that $F(\underline{x})$ is bounded below and inequality (1.6) imply that the sum

$$(2.4)\qquad \sum_{k=1}^{\infty} [F(\underline{x}^{(k)}) - F(\underline{x}^{(k+1)})]$$

is convergent. Therefore, because Σ' denotes the sum over the iterations for which condition (1.12) is satisfied, the sum

$$(2.5)\qquad \sum_{k=1}^{\infty}{}' [F(\underline{x}^{(k)}) - \Phi(\underline{x}^{(k)} + \underline{\delta}^{(k)})]$$

is convergent. Thus, by applying the elementary inequality

$$(2.6)\qquad \min[|a|, |b|] \ge \frac{|ab|}{|a| + |b|}$$

we deduce from expression (1.10) that the sum

$$(2.7)\qquad \sum_{k=1}^{\infty}{}' \frac{\|\underline{\delta}^{(k)}\| \; \|\underline{g}^{(k)}\|^2}{\|\underline{g}^{(k)}\| + \|\underline{\delta}^{(k)}\| \; \|B^{(k)}\|}$$

is also convergent.

The theorem is proved by obtaining a contradiction if $\underline{g}^{(k)}$ satisfies the bound

$$\|\underline{g}^{(k)}\| \geq c_{10}, \quad k=1,2,3,\ldots, \tag{2.8}$$

where c_{10} is a positive constant. In this case the sum (2.7) and the bounds (1.15) and (2.3) imply that the expression

$$\sum_{k=1}^{\infty}{}' \frac{\|\underline{\delta}^{(k)}\|}{1 + (\bar{\Delta}/c_{10})[c_8 + c_9 \sum_{i=1}^{k}{}' \|\underline{\delta}^{(i)}\|]} \tag{2.9}$$

is finite. Now, if $\{a_1, a_2, a_3, \ldots\}$ is any set of positive numbers such that the sum

$$\sum_{k=1}^{\infty} \left\{ a_k \Big/ \sum_{i=1}^{k} a_i \right\} \tag{2.10}$$

is convergent, then the sum Σa_k is also convergent (see Fletcher, 1972, for instance). Therefore we deduce from expression (2.9) that the sum $\Sigma' \|\underline{\delta}^{(k)}\|$ is finite. Thus inequality (2.3) shows that there is a constant c_{11} such that the bound

$$\|B^{(k)}\| \leq c_{11} \tag{2.11}$$

holds for all k. Moreover from inequality (2.1) we find the limit

$$\|\underline{\delta}^{(k)}\| \to 0. \tag{2.12}$$

By letting k_1 be an integer such that, for all $k \geq k_1$, the inequality

$$\|\underline{\delta}^{(k)}\| \leq c_{10}/c_{11} \tag{2.13}$$

is satisfied, we obtain from expressions (1.10), (2.8) and (2.11) the bound

$$F(\underline{x}^{(k)}) - \Phi(\underline{x}^{(k)} + \underline{\delta}^{(k)}) \geq c_1 \|\underline{g}^{(k)}\| \, \|\underline{\delta}^{(k)}\|, \quad k \geq k_1. \tag{2.14}$$

Thus the definition (1.3) gives the inequality

$$|1+\frac{\underline{\delta}^{(k)T}\underline{g}^{(k)}}{F(\underline{x}^{(k)})-\Phi(\underline{x}^{(k)}+\underline{\delta}^{(k)})}| = $$

(2.15)
$$|\frac{\tfrac{1}{2}\underline{\delta}^{(k)T}B^{(k)}\underline{\delta}^{(k)}}{F(\underline{x}^{(k)})-\Phi(\underline{x}^{(k)}+\underline{\delta}^{(k)})}|$$

$$\le \frac{\tfrac{1}{2}|\underline{\delta}^{(k)T}B^{(k)}\underline{\delta}^{(k)}|}{c_1\,\|\underline{g}^{(k)}\|\,\|\underline{\delta}^{(k)}\|},\quad k \ge k_1 .$$

Because expressions (2.8), (2.11) and (2.12) imply that this right-hand side tends to zero as k tends to infinity, we have the equation

$$\lim_{k\to\infty}\frac{\underline{\delta}^{(k)T}\underline{g}^{(k)}}{\Phi(\underline{x}^{(k)}+\underline{\delta}^{(k)})-F(\underline{x}^{(k)})} = 1, \tag{2.16}$$

which we are going to use to show that inequality (1.12) is satisfied for all sufficiently large values of k. We let $k_2 \ge k_1$ be an integer such that the left-hand side of equation (2.16) is at least ½ for $k \ge k_2$, and deduce from expressions (1.4) and (2.14) that the condition

$$-\underline{\delta}^{(k)T}\underline{g}^{(k)} \ge \tfrac{1}{2}[F(\underline{x}^{(k)})-\Phi(\underline{x}^{(k)}+\underline{\delta}^{(k)})] \tag{2.17}$$

$$\ge \tfrac{1}{2}\,c_1\,\|\underline{g}^{(k)}\|\,\|\underline{\delta}^{(k)}\|,\quad k \ge k_2,$$

is obtained.

To take account of the left-hand side of inequality (1.12), we require the bound

$$
\begin{aligned}
&|F(\underline{x}^{(k)}+\underline{\delta}^{(k)}) - F(\underline{x}^{(k)}) - \underline{\delta}^{(k)T}\underline{g}^{(k)}| \\
&= |\int_{\theta=0}^{1} \underline{\delta}^{(k)T}\{\underline{g}(\underline{x}^{(k)}+\theta\underline{\delta}^{(k)}) - \underline{g}^{(k)}\}d\theta| \\
&\le \|\underline{\delta}^{(k)}\| \int_{\theta=0}^{1} \|\underline{g}(\underline{x}^{(k)}+\theta\underline{\delta}^{(k)}) - \underline{g}^{(k)}\|\, d\theta \qquad (2.18) \\
&\le \|\underline{\delta}^{(k)}\|\, w(\|\underline{\delta}^{(k)}\|) ,
\end{aligned}
$$

where $w(\cdot)$ is the modulus of continuity of $\underline{g}(\underline{x})$. Thus we obtain the equation

$$
(2.19)\quad \lim_{k\to\infty} \frac{F(\underline{x}^{(k)}+\underline{\delta}^{(k)}) - F(\underline{x}^{(k)})}{\|\underline{\delta}^{(k)}\|} = \lim_{k\to\infty} \frac{\underline{\delta}^{(k)T}\underline{g}^{(k)}}{\|\underline{\delta}^{(k)}\|}
$$

Since expressions (2.8) and (2.17) show that this right hand side is bounded away from zero, the ratio of the left hand side to the right hand side of equation (2.19) tends to one. Therefore equation (2.16) gives the limit

$$
(2.20)\qquad \lim_{k\to\infty} \frac{F(\underline{x}^{(k)}) - F(\underline{x}^{(k)}+\underline{\delta}^{(k)})}{F(\underline{x}^{(k)}) - \Phi(\underline{x}^{(k)}+\underline{\delta}^{(k)})} = 1 ,
$$

showing that the test (1.12) holds for all sufficiently large k. Thus inequality (1.13) gives the condition

$$
(2.21)\qquad \Delta^{(k+1)} \ge \|\underline{\delta}^{(k)}\| , \; k \ge k_3,
$$

where k_3 is a constant integer.

Now the definition of $\delta^{(k)}$ in Section 1 is such that the value of $\|\underline{\delta}^{(k)}\|$ is either $\Delta^{(k)}$ or $\|[B^{(k)}]^{-1}\underline{g}^{(k)}\|$, and inequalities (2.8) and (2.11) provide the bound

$$
(2.22)\qquad
\begin{aligned}
\|[B^{(k)}]^{-1}\underline{g}^{(k)}\| &\ge \|\underline{g}^{(k)}\| / \|B^{(k)}\| \\
&\ge c_{10}/c_{11} ,
\end{aligned}
$$

so expression (2.12) implies that there is an integer k_4 such that the condition

(2.23) $$\|\underline{\delta}^{(k)}\| = \Delta^{(k)} \quad , \; k \geq k_4$$

is true. It follows from inequality (2.21) that the bound

(2.24) $$\|\underline{\delta}^{(k+1)}\| \geq \|\underline{\delta}^{(k)}\| \; , \; k \geq \max[k_3, k_4-1],$$

is satisfied. In other words, after a finite number of iterations, the sequence of numbers $\|\underline{\delta}^{(k)}\|$ $(k=1,2,3...)$ stops decreasing. Since inequality (1.4) implies that these numbers are positive, the limit (2.12) is not obtained. This contradiction shows that inequality (2.8) is not satisfied for any positive constant c_{10}, which proves Theorem 1.

3. Convergence to a local minimum

Now we prove that, if the sequence $\underline{x}^{(k)}$ converges to a limit point $\underline{x}^*$, if in a neighborhood of $\underline{x}^*$ second derivatives of $F(\underline{x})$ exist and are continuous, and if the matrix $G(\underline{x}^*)$ is positive definite, then the following theorem is true.

Theorem 2 The sum $\Sigma\|\underline{\delta}^{(k)}\|$ is convergent.

Proof Because, in the proof of Theorem 1, the false conjecture that $\|\underline{g}^{(k)}\|$ $(k=1,2,3,...)$ is bounded away from zero is not made until expression (2.8), inequalities (2.1) and (2.3) are still valid. However we require a result that is stronger than the statement that the sum (2.7) converges.

For this purpose we note that, if a_k $(k=1,2,3,...)$ is any sequence of monotonically decreasing positive numbers, then the sum

$$(3.1) \qquad \sum_k (a_k - a_{k+1}) / \sqrt{a_k}$$

is convergent. This result is an immeaiate consequence of the inequality

$$\sum_{k=1}^{m} (a_k - a_{k+1}) \sqrt{a_k} = \sum_{k=1}^{m} (\sqrt{a_k} - \sqrt{a_{k+1}}) (\sqrt{a_k} + \sqrt{a_{k+1}}) / \sqrt{a_k}$$

$$\leq 2 \sum_{k=1}^{m} (\sqrt{a_k} - \sqrt{a_{k+1}})$$

$$(3.2) \qquad = 2(\sqrt{a_1} - \sqrt{a_{m+1}}) < 2\sqrt{a_1} \; .$$

In particular, because the sequence $[F(\underline{x}^{(k)}) - F(\underline{x}^*)]$ $(k=1,2,3,\ldots)$ decreases monotonically, the sum

$$(3.3) \qquad \sum_{k=1}^{\infty} [F(\underline{x}^{(k)}) - F(\underline{x}^{(k+1)})] / \sqrt{F(\underline{x}^{(k)}) - F(\underline{x}^*)}$$

is convergent.

To make use of this result, we note that the conditions on $F(\underline{x})$ imply that there is a closed neighbourhood of $\underline{x}^*$, $N(\underline{x}^*)$ say, such that for all $\underline{x}$ in $N(\underline{x}^*)$ the second derivative matrix $G(\underline{x})$ is continuous and positive definite. Therefore there exist positive constants, m and M say, such that for all $\underline{y}$ and for all $\underline{x}$ in $N(\underline{x}^*)$ the bounds

$$(3.4) \qquad m \|\underline{y}\|^2 \leq \underline{y}^T G(\underline{x}) \underline{y} \leq M \|\underline{y}\|^2$$

are satisfied. We let k_5 be an integer such that $\underline{x}^{(k)}$ is in $N(\underline{x}^*)$ for all $k \geq k_5$. Then from the identity

$$(3.5) \qquad \underline{g}^{(k)} = \int_{\theta=0}^{1} G\{\underline{x}^* + \theta(\underline{x}^{(k)} - x^*)\} (\underline{x}^{(k)} - \underline{x}^*) \, d\theta ,$$

we deduce the bounds

$$(3.6) \quad m\|\underline{x}^{(k)}-\underline{x}^*\| \le \|\underline{g}^{(k)}\| \le M\|\underline{x}^{(k)}-\underline{x}^*\|, \quad k \ge k_5,$$

and from the identity

$$(3.7) \quad F(\underline{x}^{(k)}) - F(\underline{x}^*) = \int_{\theta=0}^{1} (1-\theta)(\underline{x}^{(k)}-\underline{x}^*)^T G\{\underline{x}^*+\theta(\underline{x}^{(k)}-\underline{x}^*)\}(\underline{x}^{(k)}-\underline{x}^*)d\theta$$

we deduce the bounds

$$(3.8) \quad \tfrac{1}{2}m\|\underline{x}^{(k)}-\underline{x}^*\|^2 \le F(\underline{x}^{(k)}) - F(\underline{x}^*) \le \tfrac{1}{2}M\|\underline{x}^{(k)}-\underline{x}^*\|^2, \quad k \ge k_5 .$$

It follows that the inequality

$$(3.9) \quad \sqrt{F(\underline{x}^{(k)}) - F(\underline{x}^*)} \le \|\underline{g}^{(k)}\| \sqrt{M/2m^2}, \quad k \ge k_5 ,$$

is satisfied. Therefore expression (3.3) implies that the sum

$$(3.10) \quad \sum_{k=1}^{\infty} [F(\underline{x}^{(k)}) - F(\underline{x}^{(k+1)})]/\|\underline{g}^{(k)}\|$$

is convergent.

Again we use the Σ' notation introduced in Section 2, and also inequalities (1.10) and (1.12), to deduce from expression (3.10) that the sum

$$(3.11) \quad \sum_k{}' \min[\|\underline{\delta}^{(k)}\|, \ \|\underline{g}^{(k)}\|/\|B^{(k)}\|]$$

is finite, so, by applying inequality (2.6) again, we find that the sum

$$(3.12) \quad \sum_{k=1}^{\infty}{}' \frac{\|\underline{\delta}^{(k)}\|\,\|\underline{g}^{(k)}\|}{\|\underline{g}^{(k)}\| + \|\underline{\delta}^{(k)}\|\,\|B^{(k)}\|}$$

is also convergent. To make use of this remark we note that there is a constant c_{12} such that the condition

$$(3.13) \qquad \|\underline{\delta}^{(k)}\| / \|\underline{g}^{(k)}\| \leq c_{12}, \quad k \geq k_5 ,$$

holds for all values of k that define $\underline{x}^{(k+1)}$ by the first line of equation (1.5), which includes the values of k that occur in the sum Σ'. Specifically the value of c_{12} comes from applying expressions (1.6), (3.6) and (3.8) to obtain the inequality

$$\begin{aligned}
\|\underline{\delta}^{(k)}\| &= \|\underline{x}^{(k+1)} - \underline{x}^{(k)}\| \\
&\leq \|\underline{x}^{(k)} - \underline{x}^*\| + \|\underline{x}^{(k+1)} - \underline{x}^*\| \\
&\leq \|\underline{x}^{(k)} - \underline{x}^*\| + [2\{F(\underline{x}^{(k+1)}) - F(\underline{x}^*)\}/m]^{\frac{1}{2}} \\
&\leq \|\underline{x}^{(k)} - \underline{x}^*\| + [2\{F(\underline{x}^{(k)}) - F(\underline{x}^*)\}/m]^{\frac{1}{2}} \\
&\leq (1 + \sqrt{M/m}) \; \|\underline{x}^{(k)} - \underline{x}^*\| \\
(3.14) \qquad &\leq (1 + \sqrt{M/m}) \; \|\underline{g}^{(k)}\| / m .
\end{aligned}$$

The expression (3.12) shows that the sum

$$\sum_{k=1}^{\infty}{}' \frac{\|\underline{\delta}^{(k)}\|}{1 + c_{12}\|B^{(k)}\|}$$

is finite.

It follows from inequality (2.3) and the explanation of expression (2.10) that the sum

$$(3.16) \qquad \sum_k{}' \|\underline{\delta}^{(k)}\|$$

is also finite. Therefore, because of inequality (2.1), the proof of Theorem 2 is complete.

Many of the results used in this proof are applied in the next section also. For this purpose it is convenient to change the meaning of k_5 so that, for all $k \geq k_5$, not only does $\underline{x}^{(k)}$ belong to $N(\underline{x}^*)$, but also $(\underline{x}^{(k)} + \underline{\delta}^{(k)})$ belongs to $N(\underline{x}^*)$. This change is allowed because the theorem shows that $\underline{\delta}^{(k)}$ tends to zero. Moreover we note that expressions (2.3) and (3.16) provide the bound

$$(3.17) \qquad \|B^{(k)}\| \le c_{13} ,$$

where c_{13} is a constant.

4. Superlinear convergence

If the conditions stated at the beginning of Section 3 are satisfied, then the following theorem is true for all algorithms of the class that satisfy condition (1.17).

Theorem 3 The points $\underline{x}^{(k)}$ converge to $\underline{x}^*$ superlinearly in the sense that the ratio

$$(4.1) \qquad \|\underline{x}^{(k+1)}-\underline{x}^*\| / \|\underline{x}^{(k)}-\underline{x}^*\|$$

tends to zero as k tends to infinity.

Proof First we prove that there is a positive constant c_{14} such that the inequality

$$(4.2) \qquad F(\underline{x}^{(k)}) - \Phi(\underline{x}^{(k)}+\underline{\delta}^{(k)}) \ge c_{14} \|\underline{\delta}^{(k)}\|^2$$

is satisfied. To obtain this result we show that the ratio $\|\underline{g}^{(k)}\|/\|\underline{\delta}^{(k)}\|$ is bounded away from zero, and then we apply inequality (1.10). For this purpose we let $k_6 \ge k_5$ be an integer such that the condition

$$(4.3) \qquad \|\underline{g}(\underline{x}^{(k)}+\underline{\delta}^{(k)}) - \underline{g}^{(k)} - B^{(k)}\underline{\delta}^{(k)}\| \le \tfrac{1}{2}m \|\underline{\delta}^{(k)}\| , \quad k \ge k_6 ,$$

holds, which is possible because of the limit (1.17). Also we note that the bounds (3.4) on $G(\underline{x})$ give the inequality

$$\underline{\delta}^{(k)T}[\underline{g}(\underline{x}^{(k)}+\underline{\delta}^{(k)}) - \underline{g}^{(k)}] = \int_{\theta=0}^{1} \underline{\delta}^{(k)T}\, G\{\underline{x}^{(k)}+\theta\underline{\delta}^{(k)}\}\underline{\delta}^{(k)}\, d\theta$$

$$\geq m\,\|\underline{\delta}^{(k)}\|^{2}, \quad k \geq k_5 . \tag{4.4}$$

Therefore, by applying expressions (1.4), (1.3), (4.3) and (4.4) in sequence, we deduce the bound

$$\begin{aligned} 0 &< F(\underline{x}^{(k)}) - \Phi(\underline{x}^{(k)}+\underline{\delta}^{(k)}) \\ &= -\underline{\delta}^{(k)T}\underline{g}^{(k)} - \tfrac{1}{2}\underline{\delta}^{(k)T}B^{(k)}\underline{\delta}^{(k)} \\ &\leq -\underline{\delta}^{(k)T}\underline{g}^{(k)} - \tfrac{1}{2}\underline{\delta}^{(k)T}[\underline{g}(\underline{x}^{(k)}+\underline{\delta}^{(k)})-\underline{g}^{(k)}] \\ &\qquad + \tfrac{1}{4}m\,\|\underline{\delta}^{(k)}\|^{2} \\ &\leq -\underline{\delta}^{(k)T}\underline{g}^{(k)} - \tfrac{1}{4}m\,\|\underline{\delta}^{(k)}\|^{2}, \quad k \geq k_6 , \end{aligned} \tag{4.5}$$

which gives the inequality

$$\|\underline{g}^{(k)}\| \geq \tfrac{1}{4}m\|\underline{\delta}^{(k)}\|, \quad k \geq k_6 . \tag{4.6}$$

Therefore the required bound (4.2) is an immediate consequence of expressions (1.10) and (3.17) .

Next we prove that, as k tends to infinity, the ratio

$$[F(\underline{x}^{(k)}+\underline{\delta}^{(k)}) - \Phi(\underline{x}^{(k)}+\underline{\delta}^{(k)})]/\|\underline{\delta}^{(k)}\|^{2} \tag{4.7}$$

tends to zero. We use the equation

$$F(\underline{x}^{(k)}+\underline{\delta}^{(k)}) - F(\underline{x}^{(k)}) = \tfrac{1}{2}\{\underline{g}^{(k)}+\underline{g}(\underline{x}^{(k)}+\underline{\delta}^{(k)})\}^{T}\underline{\delta}^{(k)}$$

$$+ \int_{\theta=0}^{1} (\tfrac{1}{2}-\theta)\,\underline{\delta}^{(k)T}G(\underline{x}^{(k)}+\theta\underline{\delta}^{(k)})\,\underline{\delta}^{(k)} d\theta, \tag{4.8}$$

which may be verified by integration by parts, and we note the bound

$$\left|\int_{\theta=0}^{1} (\tfrac{1}{2}-\theta)\underline{\delta}^{(k)T}\, G(\underline{x}^{(k)}+\theta\underline{\delta}^{(k)})\, \underline{\delta}^{(k)}\, d\theta\right|$$

$$= \left|\int_{\theta=0}^{1} (\tfrac{1}{2}-\theta)\underline{\delta}^{(k)T}\{G(\underline{x}^{(k)}+\theta\underline{\delta}^{(k)}) - G(x^{(k)}+\tfrac{1}{2}\underline{\delta}^{(k)})\}\, \underline{\delta}^{(k)}\, d\theta\right|$$

$$\leq \|\underline{\delta}^{(k)}\|^2 \int_{\theta=0}^{1} |\tfrac{1}{2}-\theta|\; \Omega(\{\theta-\tfrac{1}{2}\}\, \|\underline{\delta}^{(k)}\|)\, d\theta$$

$$(4.9) \quad \leq \tfrac{1}{4}\|\underline{\delta}^{(k)}\|^2\, \Omega(\tfrac{1}{2}\|\underline{\delta}^{(k)}\|), \quad k \geq k_5 ,$$

where $\Omega(\cdot)$ is the modulus of continuity of the second derivative matrix $G(\underline{x})$ in the neighbourhood $N(\underline{x}^*)$. We also require the equation

$$\Phi(\underline{x}^{(k)}+\underline{\delta}^{(k)}) - F(\underline{x}^{(k)}) = \underline{\delta}^{(k)T}\underline{g}^{(k)} + + \tfrac{1}{2}\underline{\delta}^{(k)T}B^{(k)}\underline{\delta}^{(k)}$$

$$= \tfrac{1}{2}\{\underline{g}^{(k)}+\underline{g}(\underline{x}^{(k)}+\underline{\delta}^{(k)})\}^T\,\underline{\delta}^{(k)}+$$

$$(4.10) \quad \tfrac{1}{2}\underline{\delta}^{(k)T}\,\{B^{(k)}\underline{\delta}^{(k)}+\underline{g}^{(k)}-\underline{g}(\underline{x}^{(k)}+\underline{\delta}^{(k)})\}.$$

By subtracting equation (4.10) from equation (4.8), and by using expressions (1.17), (4.9) and Theorem 2, we conclude that the ratio (4.7) does tend to zero.

This result is important because, taking account of inequality (4.2), it proves that the ratio

$$(4.11) \quad [F(\underline{x}^{(k)}) - \Phi(\underline{x}^{(k)}+\underline{\delta}^{(k)})]/[F(\underline{x}^{(k)}) - F(\underline{x}^{(k)}+\underline{\delta}^{(k)})]$$

tends to one. Thus there exists an integer, k_7 say, such that for all $k \geq k_7$ inequality (1.12) is satisfied, so the conditions

$$(4.12) \qquad \left.\begin{aligned} \Delta^{(k+1)} &\geq \|\underline{\delta}^{(k)}\| \\ \underline{x}^{(k+1)} &= \underline{x}^{(k)}+\underline{\delta}^{(k)} \end{aligned}\right\}, \; k \geq k_7 ,$$

hold. Therefore, if an iteration gives the reduction

$$(4.13) \qquad \|\underline{\delta}^{(k+1)}\| < \|\underline{\delta}^{(k)}\| ,$$

the rules governing the definition of $\underline{\delta}^{(k+1)}$ imply that it is defined by the Newton formula

$$(4.14) \qquad \underline{\delta}^{(k+1)} = -[B^{(k+1)}]^{-1} \underline{g}^{(k+1)} .$$

Since $\underline{\delta}^{(k)}$ tends to zero as k tends to infinity, it follows that the Newton formula is applied an infinite number of times.

Therefore we can find an integer, k_8 say, such that $k_8 \geq \max[k_5,k_7]$, such that $\underline{\delta}^{(k_8)}$ is defined by the Newton formula, and such that the inequality

$$(4.15) \qquad \|\underline{g}(\underline{x}^{(k)}+\underline{\delta}^{(k)}) - \underline{g}^{(k)} - B^{(k)}\underline{\delta}^{(k)}\| < \|\underline{\delta}^{(k)}\|/c_{12}, \quad k \geq k_8 ,$$

holds, where the constant c_{12} is the right hand side of expression (3.13). This last property of k_8 is obtained from the limit (1.17). Let $k \geq k_8$ be any integer such that $\underline{\delta}^{(k)}$ is the vector (1.8). Then expressions (1.8) and (4.15) and the second line (4.12) imply the bound

$$(4.16) \qquad \|\underline{g}^{(k+1)}\| < \|\underline{\delta}^{(k)}\|/c_{12} .$$

Therefore, by substituting (k+1) in place of k in inequality (3.13), we deduce that condition (4.13) is satisfied, and we recall that in this case $\underline{\delta}^{(k+1)}$ is the vector (4.14). Because we have shown that if

$k \geq k_8$ then equation (1.8) implies equation (4.14), it follows by induction that $\underline{\delta}^{(k)}$ is defined by the Newton formula for all $k \geq k_8$.

Therefore, remembering the second line of expression (4.12), the condition (1.17) gives the limit

(4.17) $$\|\underline{g}^{(k+1)}\| / \|\underline{\delta}^{(k)}\| \to 0 .$$

It follows from inequality (3.13) that the limit

(4.18) $$\|\underline{g}^{(k+1)}\| / \|\underline{g}^{(k)}\| \to 0$$

is obtained, so the bounds (3.6) imply that the ratio (4.1) tends to zero. Theorem 3 is proved.

5. Applications

In this section we note that our theory applies to a number of useful algorithms, mainly because condition (1.10) is quite natural. This fact is shown by the following theorem.

Theorem 4 Let $\underline{\eta}^{(k)}$ be the value of $\underline{\delta}$ that minimizes $\Phi(\underline{x}^{(k)}+\underline{\delta})$, subject to the inequality

(5.1) $$\|\underline{\eta}^{(k)}\| \leq \|\underline{\delta}^{(k)}\| ,$$

and subject to the condition that $\underline{\eta}^{(k)}$ has the form

(5.2) $$\underline{\eta}^{(k)} = -\alpha \underline{g}^{(k)} .$$

Then the bound

(5.3) $$F(\underline{x}^{(k)}) - \Phi(\underline{x}^{(k)}+\underline{\eta}^{(k)}) \geq \tfrac{1}{2} \|\underline{g}^{(k)}\| \min[\, \|\underline{\delta}^{(k)}\| , \|\underline{g}^{(k)}\| / \|B^{(k)}\| \,]$$

is obtained.

Proof The definition (1.3) provides the equation

$$(5.4) \quad F(\underline{x}^{(k)}) - \Phi(\underline{x}^{(k)}+\underline{\eta}^{(k)}) = \alpha\|\underline{g}^{(k)}\|^2 - \tfrac{1}{2}\alpha^2\, \underline{g}^{(k)T} B^{(k)} \underline{g}^{(k)} .$$

Therefore, if the condition

$$(5.5) \quad \underline{g}^{(k)T} B^{(k)} \underline{g}^{(k)} \le 0$$

holds, then the required vector $\underline{\eta}^{(k)}$ is obtained when α has the value

$$(5.6) \quad \alpha = \|\underline{\delta}^{(k)}\| / \|\underline{g}^{(k)}\| ,$$

in which case the inequality

$$(5.7) \quad F(\underline{x}^{(k)}) - \Phi(\underline{x}^{(k)}+\underline{\eta}^{(k)}) \ge \|\underline{\delta}^{(k)}\| \; \|\underline{g}^{(k)}\|$$

is satisfied, which is consistent with the bound (5.3). However, if condition (5.5) is not obtained, then α is the number

$$(5.8) \quad \alpha = \min[\|\underline{\delta}^{(k)}\| / \|\underline{g}^{(k)}\| , \; \|\underline{g}^{(k)}\|^2 / \underline{g}^{(k)T} B^{(k)} \underline{g}^{(k)}] \ge \min[\|\underline{\delta}^{(k)}\| / \|\underline{g}^{(k)}\| , \; 1/\|B^{(k)}\|].$$

In this case the fact that expression (5.4) is quadratic in α provides the bound

$$(5.9) \quad F(\underline{x}^{(k)}) - \Phi(\underline{x}^{(k)}+\underline{\eta}^{(k)}) \ge \tfrac{1}{2}\alpha \|\underline{g}^{(k)}\|^2 \ge \tfrac{1}{2} \|\underline{g}^{(k)}\| \min[\|\underline{\delta}^{(k)}\| , \|\underline{g}^{(k)}\| / \|B^{(k)}\|],$$

which is also consistent with condition (5.3). Theorem 4 is proved.

This theorem shows that we can view condition (1.10) in the following way. We require the difference

$$(5.10) \quad F(\underline{x}^{(k)}) - \Phi(\underline{x}^{(k)} + \underline{\delta}^{(k)})$$

to be no less than a constant positive multiple of the greatest value of the difference

$$F(\underline{x}^{(k)}) - \Phi(\underline{x}^{(k)} + \underline{\eta}^{(k)}) \tag{5.11}$$

that can be obtained when $\underline{\eta}^{(k)}$ is subject to conditions (5.1) and (5.2). This property is obtained by many applications of the Marquardt technique, see Goldfeld, Quandt and Trotter (1966) and Hebden (1973) for instance, by the dog-leg method (Powell, 1970), by the unconstrained version of the hypercube algorithm (Fletcher, 1972) and by the Newton formula (1.8) when the matrix $B^{(k)}$ is positive definite. Also, because the steepest descent algorithm can be derived by setting $B^{(k)} = 0$ in the approximation (1.3), the test (1.10) allows a practical version of the steepest descent method. However condition (1.10) and the method for controlling the length of $\underline{\delta}^{(k)}$ rule out all current versions of variable metric algorithms.

Condition (1.11) is also satisfied by a number of practical algorithms. The most obvious examples are those for which the matrices $B^{(k)}$ are bounded explicitly, including methods that set $B^{(k)} = G(\underline{x}^{(k)})$ when $G(\underline{x})$ is uniformly bounded, and methods that use the Gauss technique to minimize a sum of squares of nonlinear functions when first derivatives are bounded (see Kowalik and Osborne, 1968, for instance). In these cases the sophistication of the proofs of Theorems 1 and 2 is unnecessary because c_3 is zero in inequality (1.11). However the term $c_3 \sum_i \|\underline{\delta}^{(i)}\|$ is useful because it admits the updating formula

$$B^{(k+1)} = B^{(k)} + \frac{\underline{\sigma}^{(k)}\underline{\delta}^{(k)T} + \underline{\delta}^{(k)}\underline{\sigma}^{(k)T}}{||\underline{\delta}^{(k)}||^2} - \frac{\underline{\delta}^{(k)}\underline{\delta}^{(k)T}(\underline{\sigma}^{(k)T}\underline{\delta}^{(k)})}{||\underline{\delta}^{(k)}||^4}, \tag{5.12}$$

where $\underline{\sigma}^{(k)}$ is the vector

$$\underline{\sigma}^{(k)} = \underline{g}(\underline{x}^{(k)}+\underline{\delta}^{(k)}) - \underline{g}^{(k)} - B^{(k)}\underline{\delta}^{(k)} . \tag{5.13}$$

This formula is applied by Powell (1970) and by Fletcher (1972), and proofs that it satisfies condition (1.11) are given by Fletcher (1972) and by Dennis (1972), which depend on $G(\underline{x})$ satisfying the Lipschitz condition

$$||G(\underline{x}) - G(\underline{y})|| \leq L\,||\underline{x}-\underline{y}|| \tag{5.14}$$

for all $\underline{x}$ and $\underline{y}$, where L is a constant.

Algorithms of our class that apply formula (5.12) usually have superlinear convergence, because they satisfy condition (1.17), which makes Theorem 3 valid. We now prove this statement.

Theorem 5 If the conditions stated at the beginning of Section 3 are satisfied, and if the Lipschitz condition (5.14) holds for all $\underline{x}$ and $\underline{y}$ in a neighbourhood of $\underline{x}^*$, then the limit (1.17) is obtained by all algorithms of our class that calculate the sequence of matrices $B^{(k)}$ $(k=1,2,3,\ldots)$ from formula (5.12).

Proof If $\bar{G}^{(k)}$ is any matrix that satisfies the condition

$$\underline{g}(\underline{x}^{(k)}+\underline{\delta}^{(k)}) - \underline{g}^{(k)} = \bar{G}^{(k)}\underline{\delta}^{(k)} , \tag{5.15}$$

then it follows from formula (5.12) that the equation

$$(5.16)\quad (B^{(k+1)}-\bar{G}^{(k)}) = \left(I - \frac{\underline{\delta}^{(k)}\underline{\delta}^{(k)T}}{\|\underline{\delta}^{(k)}\|^2}\right)(B^{(k)}-\bar{G}^{(k)})\left(I - \frac{\underline{\delta}^{(k)}\underline{\delta}^{(k)T}}{\|\underline{\delta}^{(k)}\|^2}\right)$$

holds. It is convenient to introduce Euclidean matrix norms, which we denote by the subscript E, while all unsubscripted matrix norms remain subordinate to the Euclidean vector norm. Since the Euclidean norm of a matrix is the square root of the sum of squares of the elements of the matrix, the inequality

$$(5.17)\qquad \|A\|_E \leq \sqrt{n}\ \|A\|$$

holds for all $n \times n$ matrices A.

The proof depends on showing that the sum

$$(5.18)\qquad \sum_{k=k_5}^{m} \|B^{(k)} - \bar{G}^{(k)}\|_E^2 - \|B^{(k+1)} - \bar{G}^{(k)}\|_E^2$$

is bounded above for all $m > k_5$, where k_5 is the integer discussed at the end of Section 3, and where $\bar{G}^{(k)}$ is the matrix

$$(5.19)\qquad \bar{G}^{(k)} = \int_{\theta=0}^{1} G(\underline{x}^{(k)} + \theta\underline{\delta}^{(k)})\, d\theta ,$$

which satisfies condition (5.15). We make use of the fact that for all $n \times n$ matrices A, B and C the inequality

$$\|A-B\|_E^2 - \|A-C\|_E^2 = [\|A-B\|_E - \|A-C\|_E]\,[\|A-B\|_E + \|A-C\|_E]$$

$$(5.20)\qquad \leq \|B-C\|_E\,[2\,\|A\|_E + \|B\|_E + \|C\|_E]$$

is satisfied. Moreover we note that the conditions on $F(\underline{x})$ given in Section 3 imply that there is a constant c_{15} such that the inequalities

$$\left.\begin{aligned} \|G^{(k)}\| &\le c_{15} \\ \|\bar{G}^{(k)}\| &\le c_{15} \end{aligned}\right\}, \quad k \ge k_5, \tag{5.21}$$

hold. Also from the Lipschitz condition (5.14) we obtain the bound

$$\begin{aligned} \|\bar{G}^{(k)} - G(\underline{x}^{(k)})\| &= \left\|\int_{\theta=0}^{1} [G(\underline{x}^{(k)}+\theta\underline{\delta}^{(k)}) - G(\underline{x}^{(k)})]\, d\theta\right\| \\ &\le \tfrac{1}{2}L\, \|\underline{\delta}^{(k)}\| \quad , \end{aligned} \tag{5.22}$$

and similarly we deduce the condition

$$\|\bar{G}^{(k)} - G(\underline{x}^{(k+1)})\| \le \tfrac{1}{2}L \;\; \|\underline{\delta}^{(k)}\| \;. \tag{5.23}$$

From expressions (5.20), (5.17), (5.22), (3.17) and (5.21) we infer the inequality

$$\begin{aligned} \|B^{(k)} - \bar{G}^{(k)}\|_E^2 &- \|B^{(k)} - G^{(k)}\|_E^2 \\ &\le n\,\|\bar{G}^{(k)} - G^{(k)}\| \;\; [2\|B^{(k)}\| + \|\bar{G}^{(k)}\| + \|G^{(k)}\|] \\ &\le n\,L(c_{13}+c_{15}) \;\; \|\underline{\delta}^{(k)}\|, \quad k \ge k_5 \,, \end{aligned} \tag{5.24}$$

and similarly we obtain the bound

$$\begin{aligned} \|B^{(k+1)} - G^{(k+1)}\|_E^2 &- \|B^{(k+1)} - \bar{G}^{(k)}\|_E^2 \\ &\le n\,L(c_{13}+c_{15}) \;\; \|\underline{\delta}^{(k)}\|, \quad k \ge k_5 \,. \end{aligned} \tag{5.25}$$

Thus the sum (5.18) is bounded above by the expression

$$\begin{aligned} \sum_{k=k_5}^{m} 2nL(c_{13}+c_{15}) \;\; \|\underline{\delta}^{(k)}\| &- \|B^{(m+1)} - G^{(m+1)}\|_E^2 \\ &+ \|B^{(k_5)} - G^{(k_5)}\|_E^2 \,, \end{aligned} \tag{5.26}$$

which is finite because of Theorem 2 and inequalities (3.17) and (5.21). This remark is helpful, because the definition of the Euclidean matrix norm and equation (5.16) give the inequality

$$(5.26) \qquad \|B^{(k)}-\bar{G}^{(k)}\|_E^2 - \|B^{(k+1)}-\bar{G}^{(k)}\|_E^2$$

$$\geq \|(B^{(k)}-\bar{G}^{(k)}\|_E^2 - \left\|(B^{(k)}-\bar{G}^{(k)})\left(I - \frac{\underline{\delta}^{(k)}\underline{\delta}^{(k)T}}{\|\underline{\delta}^{(k)}\|^2}\right)\right\|_E^2$$

$$= \|(B^{(k)}-\bar{G}^{(k)})\,\underline{\delta}^{(k)}\|^2 / \|\underline{\delta}^{(k)}\|^2$$

$$= \|B^{(k)}\underline{\delta}^{(k)} + \underline{g}^{(k)} - \underline{g}(\underline{x}^{(k)}+\underline{\delta}^{(k)})\|^2 / \|\underline{\delta}^{(k)}\|^2,$$

the last line being a consequence of equation (5.15). Therefore, because we have shown that the sum (5.18) is bounded above, it follows that the sum

$$(5.27) \qquad \sum_k \|B^{(k)}\underline{\delta}^{(k)} + \underline{g}^{(k)} - \underline{g}(\underline{x}^{(k)} + \underline{\delta}^{(k)})\|^2 / \|\underline{\delta}^{(k)}\|^2$$

is convergent, which implies the limit (1.17). Theorem 5 is proved.

Although many minimization algorithms obtain superlinear convergence by forcing the errors of second derivative approximations to zero, note that Theorems 3 and 5 do not imply that $B^{(k)}$ converges to $G(\underline{x}^*)$. An example given by Dennis and Moré shows that it is not necessary for $B^{(k)}$ to converge to $G(\underline{x}^*)$ as k tends to infinity.

Mr. H. Ramsin of the Royal Institute of Technology, Stockholm is investigating whether Theorem 1 remains valid if inequality (1.11) is replaced by the weaker condition

$$(5.28) \qquad \|B^{(k)}\| \leq c_{16} + kc_{17},$$

where c_{16} and c_{17} are positive constants. I believe that this conjecture is true. In this case Theorem 1 holds for the updating formula (5.12) under less restrictive conditions on $F(\underline{x})$. Specifically we may replace the Lipschitz condition (5.14) on the second derivative of $F(\underline{x})$, by a similar Lipschitz condition on the first derivative vector $\underline{g}(\underline{x})$.

References

Dennis, J.E. (1972) "On some methods based on Broyden's secant approximation to the Hessian", in "Numerical methods for nonlinear optimization", ed. F.A. Lootsma, Academic Press.

Dennis, J.E. and Moré, J.J. (1973) "A characterization of superlinear convergence and its application to quasi-Newton methods", Report TR 73-157, Cornell University.

Fletcher, R. (1972) "An algorithm for solving linearly constrained optimization problems", Math. Prog., Vol. 2, pp. 133-165.

Goldfeld, S.M., Quandt, R.E. and Trotter, H.F. (1966) "Maximization by quadratic hill-climbing", Econometrica, Vol. 34, pp. 541-551.

Hebden, M.D. (1973) "An algorithm for minimization using exact second derivatives", Report TP.515, A.E.R.E. Harwell.

Kowalik, J. and Osborne, M.R. (1968) "Methods for unconstrained optimization problems", Elsevier Inc.

Marquardt, D.W. (1963) "An algorithm for least squares estimation of nonlinear parameters", SIAM Journal, Vol. 11, pp. 431-441.

Powell, M.J.D. (1970) "A new algorithm for unconstrained optimization" in "Nonlinear programming", eds. J.B. Rosen, O.L. Mangasarian and K. Ritter, Academic Press.

CONVERGENCE OF THE REDUCED GRADIENT METHOD

by

Pierre Huard
Electricité de France

1. Introduction

The Reduced Gradient method is not new, seeing that it was proposed by Ph. WOLFE in 1962 [4], [1]. Although its field of application is limited to the solving of linear constraints and continuously differentiable nonlinear objective function programmes, it has successfully been used since that date to solve real problems of large size. It has not, however, been possible to establish its convergence to an optimal solution. Fortunately, it may be said, for in 1966 its author published a simple numerical example for which the algorithm converged to a non-optimal solution [5].

The Reduced Gradient method includes operations similar to those of the Simplex method (in particular the use of bases and pivotal operations), with additional operations resulting from the non-linearity of the objective function (computation of its derivatives at each new point, maximization of this function on a segment). Since the rules to be followed in the original method, during pivotal operations, offer a wide margin of freedom, an interesting theoretical problem consisted in the quest for more precise rules to ensure convergence of the method.

The present paper partly answers the question, insofar as it describes a variant of the Reduced Gradient method providing a sequence, generally infinite, of feasible solutions whose accumulation points are optimal solutions of the problem posed. In the case of a linear objective function, the algorithm proposed is considerably simplified, and converges in a finite number of iterations. But the demonstration is based on a very strong assumption of non-degeneracy of feasible solutions (hypothesis (H2) of section 1).

At the end of this paper (section 6), it is shown that a classical polynomial disturbance of the second member of the constraints, as small as one may wish, transforms the problem set into a neighboring problem that verifies the non-degeneracy hypothesis. It would be interesting to get entirely rid of this hypothesis without modifying the problem, by applying a rule similar to the lexicographic rule of the Simplex method, but the difficulties inherent in non-extreme feasible solutions are present in the Reduced Gradient method.

The algorithm is described in detail. It will, however, be easier to read if one already has a nodding acquaintance with the original Reduced Gradient, for instance by reading the articles quoted : some of the results established in these articles are used here without any demonstration.

We also assume known the theoretical method of the Gradient Projection [3], [2], which permits maximizing a continuously differentiable concave function, on a closed convex set, by successive projections of its gradient : the development of the Reduced Gradient often becomes identified, after a finite number of pivotal operations, with that of the Gradient Projection. But we have failed to establish that this eventuality always occurs.

What characterizes the variant of the Reduced method presented here consists essentially in three points. Firstly, the variables leaving the basis are "provisionally forbidden candidacy". Furthermore, candidates for entry into the basis are chosen only from among the strictly positive variables. Lastly, if during the course of certain iterations, the movement is "too small", the derivatives are not recalculated, in order that the Reduced Gradient, by becoming identified with the Projected Gradient, shall have its convergence ensured.

2. The Problem Set - Notation - Hypotheses

2.1 - The problem set

We consider the following Programming problem :

$$(P) \quad \begin{array}{|c|}\hline \text{Maximize } f(x) \text{ subject to} \\ A\,x = a \qquad (1) \\ x \geq 0 \qquad (2) \\ \hline \end{array}$$

where $f : R^n \to R$ a concave, twice continuously differentiable function.

A : an $(L \times J)$ - matrix, with $|L| = m$ and $|J| = n$

x, a : two columns whose sets of subscripts are respectively J and L .

2.2 - Notation

- P domain of (P) .
- The same symbol x is used for the point $x \in R^n$ and its representative column. As well as for $a \in R^m$.
- $x \cdot y$ scalar product of x by y .
- $\nabla f(x)$ and $H(x)$ respectively the gradient and the matrix of the second derivatives of f , with values at x .
- A_i, A^j, A_i^j respectively row i , column j , element (i,j) of A .
- If $L' \subset L$, $J' \subset J$, then $A_{L'}^{J'}$ is the sub-matrix of A , made up of elements A_ℓ^j , $(\ell, j) \in L' \times J'$.
- I is a basis if $I \subset J$, $|I| = m$ and rank $(A^I)^{-1} = m$.
- $\bar{I} = J - I$ (complementary set of I) .
- $\nabla_I f(x)$ a vector whose components are $\partial f(x)/\partial x_j$, $j \in I$.
- $T(I) = (A^I)^{-1} A$, $t(I) = (A^I)^{-1} a$ (Simplex Tableau and its right-hand member).
- $d(I,x) = \nabla f(x) - \nabla_I f(x) T(I)$ (Reduced gradient) (We shall write $d(I)$ instead of $d(I,x)$ if f is linear).
- To simplify, T will sometimes be written for $T(I)$. The same for t and d .

$$y_{\bar{I}}(d,x) : \left\{\begin{array}{l} y_j = 0 \quad \text{if} \quad x_j = 0 \\ \qquad \text{and} \quad d^j < 0 \\ \\ y_j = d^j \quad \text{otherwise.} \end{array}\right\} \quad \forall j \in \bar{I}$$

$$y_I(d,x) = - T^{\bar{I}}(I) \; y_{\bar{I}}(d,x)$$

$$y(d,x) = \begin{bmatrix} y_I(d,x) \\ y_{\bar{I}}(d,x) \end{bmatrix}$$ (we always have $Ay(d,x) = 0, \forall d, \forall x$)

2.3. - Hypotheses

(H_0) rank $A = m$

(H_1) P is bounded. Therefore P is compact, and f attains its maximum value on P at a point $\hat{x}$.
(If f is a linear function, this hypothesis can be replaced by : f attains its maximum value on P at a point $\hat{x}$).

(H_2) Setting $S(x) = \{j \in J \mid x_j \neq 0\}$, we have :
$x \in P \Longrightarrow$ rank $A^{S(x)} = m$
(and therefore $|S(x)| \geq m$).

(H_3) $\exists \beta' > 0 : - \beta' |y|^2 \leq y.H(x)y \leq 0$,
$\forall x \in R^n$, $\forall y \in R^n$
(f being a concave function, $H(x)$ is a negative semidefinite matrix).

3. Lemmas (Classical results)

Let $x \in P$ and I be a basis.

3.1. $y_{\bar{I}}(d(I,x),x) = 0 \Longrightarrow x$ optimal solution of (P).

3.2. If $\exists d \in R^m$ such that :

$y(d,x) \geq 0$ and $\neq 0$

then P is not a bounded set.

If furthermore f is linear and $d = d(I)$, then (P) has no finite optimal solution.

3.3 Putting $R = \{i \in I \mid i \notin S(x)\}$

$S' = \bar{I} \cap S(x)$

if $S' = \emptyset$ and rank $A^{S(x)} = m$, then

$\forall r \in R, \ \exists s \in S' : T_r^S(I) \neq 0$

Proof:

1) cf [1] page 172. Under the above conditions, we have $d(I,x) \leq 0$ and $d(I,x).x = 0$, which gives the Kuhn-Tucker sufficient optimality conditions, with $- \nabla_I f(x). \ (A^I)^{-1}$ as multiplier.

2) We verify that $x + \theta y \in P, \ \forall \theta \geq 0$, and if f is linear and $d = d(I)$:

$$f(x+\theta y) = f(x) + \theta |y_{\bar{I}}|^2 \to + \infty \text{ with } \theta$$

3) Writing S instead of S(x), T instead of T(I) to simplify, we have:

$$\left.\begin{array}{l} \left.\begin{array}{l} \text{rank } A^S = m \\ \text{rank}(A^I)^{-1} = m \end{array}\right\} \Longrightarrow \text{rank } T^S = m \\ R \subset I \Longrightarrow T_R^{I-R} = 0 \end{array}\right\} \Longrightarrow T_r^{S'} \neq 0, \forall r \in R$$

4. Algorithm

4.1. General case (f nonlinear)

Start with $\overset{o}{x} \in P$ and any basis I_o.

Set $E_o = \emptyset$

Step k

① (Up dating the values)

We have $\overset{k}{x} \in P$, I_k a basis,

$$\bar{I}_k = J - I_k$$

$$E_k \subset \bar{I}_k \quad \text{(the set of forbidden candidates).}$$

Determine $T = T(I_k)$, $t = t(I_k)$

$$d_k = d(I_k, \overset{k}{x}), \quad S_k = S(\overset{k}{x}) \; .$$

Set $h = 1$, $\overset{k1}{z} = \overset{k}{x}$, $\overset{k1}{y} = y(d_k, \overset{k}{x})$

1.1. If $\overset{k1}{y} = 0$, $\overset{k}{x}$ is an optimal solution of (P). Stop.

1.2. Otherwise, go to ② .

② (Determination of the next solution)

We have $\overset{kh}{z}$ and $\overset{kh}{y} \neq 0$

Determine $\overset{k(h+1)}{z} = \overset{kh}{z} + \theta_{kh} \overset{kh}{y}$

with $\theta_{kh} = \min \{ \overset{kh}{z}_j / -\overset{kh}{y}_j \mid j \in J, \overset{kh}{y}_j < 0 \}$

Set $R_{kh} = \{ j \in J \mid \theta_{kh} = \overset{kh}{z}_j / -\overset{kh}{y}_j , \overset{kh}{y}_j < 0 \}$

Go to ③

<u>Remarks</u>: $R_{kh} \neq \emptyset$, since $\overset{kh}{y} \neq 0$, if not, with lemma 3.2, P would not be bounded (resp. if f is linear: P would not have a finite optimal solution), in contradiction with (H1) .

- $\overset{k(h+1)}{z_j} = 0, \forall j \in R_{kh}$

(3) (Stopping the projection of d_k)

3.1 If $h = 1$ and $\theta_{k1} = 0$, go to (5)

<u>Remark</u>: $R_{k1} \cap \bar{I}_k = \emptyset$, then $R_{k1} \subset I_k$

3.2 If $h > 1$ and $\theta_{kh} = 0$, go to (6)

<u>Remark</u>: $R_{kh} \cap \bar{I}_k = \emptyset$, then $R_{kh} \subset I_k$

3.3 Otherwise, go to (4) .

(4) (Piece - wise projection of d_k)

4.1 If $\sum_{i=1}^{h} \theta_{ki} \geq 1$, modify θ_{kh} such that $\sum_{i=1}^{h} \theta_{ki} = 1$ and consequently modify $\overset{kh}{z}$. Go to (6)

4.2 Otherwise, determine:

$$\overset{k(h+1)}{y} = y(d_k, \overset{k(h+1)}{z})$$

4.2.1 If $\overset{k(h+1)}{y} = 0$, go to (6)

4.2.2 Otherwise, go to (2) , with h + 1 instead of h (k unchanged).

⑤ (Pivotal operation)

Choose $r \in R_{k1}$ and put $I_{k+1} = I_k - r + s$, with s such that:

5.1 If possible $s \in (\bar{I}_k - E_k) \cap S_k$, with $T_r^s \neq 0$ and put $E_{k+1} = E_k \cup \{r\}$

5.2 Otherwise $s \in \bar{I}_k \cap S_k$ with $T_r^s \neq 0$, and put $E_{k+1} = \{r\}$

Go to ① with $k + 1$ instead of k.

Remark: $\bar{I}_k \cap S_k \neq \emptyset$, since here $R_{k1} \subset I_k$, and from the remarks under ② and the hypothesis (H_2). Then with lemma 3.3, $\exists s \in \bar{I}_k \cap S_k$: $T_r^s \neq 0$.

⑥ (Maximizing f on a segment)

Set $\overset{k}{z} = \overset{k(h+1)}{z}$. Determine $\overset{k+1}{x}$ such that:

$$\overset{k+1}{x} \text{ maximizes } f \text{ on } \left[\overset{k}{x}, \overset{k}{z}\right]$$

Set $I_{k+1} = I_k$, $E_{k+1} = E_k$. Go to ① with $k + 1$ instead of k.

Remark: If f is linear, we always have $x^{k+1} = z^k$, for then

$$f(z^{k(i+1)}) = f(z^{ki}) + \theta_{ki} \left| y^{ki}_{\bar{I}_k} \right|^2 ,$$

$\forall i = 1,2,\ldots,h$. Therefore, if $z^k \neq x^k$,

$\exists i$ such that $\theta_{ki} > 0$ and $y^{ki} \neq 0$,

and then $f(z^k) > f(x), \forall x \in [x^k, z^k[$.

4.2 Linear case (f linear)

The preceding algorithm is appreciably simplified, because the maximization on $[x^k, z^k]$ disappears and the intermediate steps (kh) are no longer distinguishable from the general steps (k).

Start with $x^o \in P$, and any basis I_o.

Put $E_o = \emptyset$.

Step k

(1) (Updating the values)

We have $x^k \in P$, I_k a basis, $\bar{I}_k = J - I_k$

$E_k \subset \bar{I}_k$ (the set of forbidden candidates)

Determine $T = T(I_k)$, $t = t(I_k)$

$$d_k = d(I_k), \quad S_k = S(x^k)$$

$$y^k = y(d_k, x^k)$$

If $y^k = 0$, x^k is an optimal solution of (P). Stop.
Otherwise, go to (2).

(2) (Determination of the next solution)

Determine $x^{k+1} = x^k + \theta_k y^k$

with $\theta_k = \min \{x_j^k/-y_j^k \mid j \in J, y_j^k < 0\}$

Set $R_k = \{j \in J \mid \theta_k = x_j^k/-y_j^k, y_j^k < 0\}$

Go to (3)

(3) (Stopping the projection of d_k)

If $\theta_k = 0$, go to (5)

Otherwise, put $I_{k+1} = I_k$, $E_{k+1} = E_k$.

Go to (1) with k+1 instead of k.

(4) Cancelled.

(5) (Pivotal operation)
Unchanged, reading R_k instead of R_{k1}.

(6) Cancelled.

5. Convergence

With the hypotheses (H_o) to (H_3), the algorithm generates a generally infinite sequence of feasible solutions x^k such that :

- $f(x^k) \leq f(x^{k+1}) \leq f(\hat{x})$

- Every accumulation point of the infinite sequence $\{x^k | k \in N\}$ is an optimal solution of (P).

- If f is linear, then the sequence is finite, i.e. there exists an integer k such that: x^k is an optimal solution of (P).

5.1 Both the inequalities are evident, since x^{k+1} maximizes f on $[x^k, z^k] \subset P$, $\forall k \in N$.

5.2 The algorithm can be finite only if we have at some step $y^{kl}_{\bar{I}k} = 0$, and then the solution x^k is an optimal one. We shall therefore assume in what follows that this contingency never arises, and that the number of steps is infinite. Under these conditions, there are two possible cases:

(i) an infinite subsequence, defined by $N' \subset N$ such that:

$$\{ |y^{kl}_{\bar{I}k}| \; | k \in N' \} \to 0$$

(ii) $\exists \alpha > 0 : |y^{kl}_{\bar{I}k}| \geq \alpha, \; \forall k \in N$

We shall prove that in case (i) every accumulation point is an optimal solution of (P) (this result is already established in [1], pages 198 to 203). In case (ii), the number of pivotal operations is finite (mainly because of the use of the set E_k and the hypothesis (H_2)), and after the last pivotal operation, the algorithm is reduced to solving a reduced problem by means of a classical Gradient Projection method. Consequently, every accumulation point is an optimal solution.

In what follows, it will be helpful to note that for any given basis I:

$$Ax = a \Longleftrightarrow x_I = t(I) - T^{\bar{I}}(I)x_{\bar{I}}$$

and to put:

$$x_I(x_{\bar{I}}) = t(I) - T^{\bar{I}}(I)x_{\bar{I}}$$

$$f'(x_{\bar{I}}) = f(x_I(x_{\bar{I}}), x_{\bar{I}}) \quad (f' \text{ concave})$$

We then verify that for every point satisfying $Ax = a$, we can substitute $f'(x_{\bar{I}})$ for $f(x)$, the gradient being:

$$\nabla f'(x_{\bar{I}}) = d^{\bar{I}}(I,x)$$

5.2.1 <u>Hypothesis</u>:

$$\exists N' \subset N: \{|y^{kl}_{\bar{I}k}| \; |k \in N'\} \to 0$$

(This case is excluded if f is a linear function, as shown further on in section 5.3).

Noting that $Ax^k = a$ and that f' is a concave function, we have, $\forall k \in N$:

$$0 \leq f(\hat{x}) - f(\overset{k}{x}) = f'(\hat{x}_{\bar{I}k}) - f'(\overset{k}{x}_{\bar{I}k})$$

$$\leq d_k^{\bar{I}k} \cdot (\hat{x}_{\bar{I}k} - \overset{k}{x}_{\bar{I}k})$$

$$\leq \overset{kl}{y}_{\bar{I}k} \cdot (\hat{x}_{\bar{I}k} - \overset{k}{x}_{\bar{I}k})$$

the last inequality resulting from $\hat{x} \geq 0$ and

$$\overset{kl}{y}_j \neq d_k^j \iff \overset{k}{x}_j = 0 \text{ and } d_k^j < 0, \forall j \in \bar{I}_k$$

The domain being bounded, and the sequence of the values $f(\overset{k}{x})$ being monotonically nondecreasing, we then have:

$$\{f(\overset{k}{x}) \mid k \in N\} \to f(\hat{x})$$

Finally, P being compact, the sequence $\{\overset{k}{x} \in P \mid k \in N\}$ has a nonempty set of accumulation points, each of them being a feasible, and therefore optimal, solution.

5.2.2 Hypothesis

$$\exists \alpha > 0 : |\overset{kl}{y}_{\bar{I}k}| \geq \alpha, \ \forall k \in N$$

We shall first consider the case of a finite number of pivotal operations, then the case of an infinite number. In the latter, we shall prove that it is not possible.

5.2.2.1. Finite number of pivotal operations

In other words:

$$\exists k' : \left\{ \begin{array}{l} I_k = I = c^{te} \\ x_I^k \geq 0, \forall i \in I \end{array} \right\} \forall k \geq k'$$

Let us consider the reduced Programme Q(I), with variable $x_{\bar{I}}$, and equivalent to (P):

$$Q(I) \quad \boxed{\begin{array}{ll} \text{Maximise } f'(x_{\bar{I}}) \text{ subject to:} & \\ \qquad t(I) - T^{\bar{I}}(I)\, x_{\bar{I}} \geq 0 & (3) \\ \qquad x_{\bar{I}} \geq 0 & (4) \end{array}}$$

where the gradient of the concave function $f'(x_{\bar{I}})$ is $d^{\bar{I}}(I,x)$.

The constraint (3), whose left-hand side represents the values of the basic variables, is always satisfied by x^k , $\forall k \geq k'$, under our hypothesis 5.2.2.1. The algorithm leads then to the use of the Gradient Projection method for Q(I), where the constraint (3) is omitted. As a matter of fact, $z_{\bar{I}}^k$ is simply the projection of the point $x_{\bar{I}}^k + \nabla f'(x_{\bar{I}}^k)$ on the closed convex set C defined by $x_{\bar{I}} \geq 0$.

From the classical theory of the Gradient Projection method, we know that every accumulation point of $\{x^k_{\bar{I}} \mid k \in N\}$ is an optimal solution of $Q(I)$. We thus have:

$$\exists N' \subset N: \{x^k_{\bar{I}} \mid k \in N'\} \to \overset{*}{x}_{\bar{I}} \text{ optimal solution of } Q(I)$$

and then, by continuity of $x_I(x_{\bar{I}})$:

$$\{x^k \mid k \in N'\} \to \overset{*}{x} \quad \text{optimal solution of (P).}$$

<u>Remark</u>:

The theory of the Gradient Projection method gives us the following additional result:

$$\{(z^k - x^k) \mid k \in N'\} \to 0$$

On the other hand, we have:

$$z^k - x^k = \sum_{h=1}^{pk} \theta_{kh}\, y^{kh}$$

These relations do not contradict *a priori* the hypothesis 5.2.2., i.e.:

$$|y^{k1}_{\bar{I}k}| \geq \alpha > 0, \ \forall k \in N$$

because we may have $\theta_{k1} \to 0$ when $k \to \infty$, $k \in N'$.

5.2.2.2. Infinite number of pivotal operations

We shall show further on that this hypothesis is not possible. But to do this we first have to prove that the sequence of points x^k is convergent.

i) The infinite sequence $\{x^k \mid k \in N\} \to \overset{*}{x} \in P$

We put $\phi_k(\theta) = f((1-\theta)x^k + \theta z^k) - f(x^k)$

$$= f'((1-\theta)x^k_{\bar{I}k} + \theta z^k_{\bar{I}k}) - f(x^k)$$

and $\quad y^k = z^k - x^k$.

ϕ_k is a concave function, and from (H_3), it is underbounded by a second degree concave function, whose first and second derivatives at $\theta = 0$ are equal respectively to $\nabla f(x^k) \cdot y^k$ and $\beta' |y^k|^2$. Hence, using a classical result (cf. for example [1] page 200), and setting:

$$\phi_k(\hat{\theta}) = \max\{\phi_k(\theta) \mid \theta \geq 0\}$$

we have:

$$f(x^{k+1}) - f(x^k) = \max\{\phi_k(\theta) \mid \theta \in [0,1]\}\ldots$$

$$\ldots \geq \begin{cases} \frac{1}{2}(\nabla f(x^k).y^k)^2/\beta'|y^k|^2 & \text{if } \hat{\theta} \leq 1 \\ \frac{1}{2}\nabla f(x^k).y^k & \text{if } \hat{\theta} \geq 1 \end{cases} \tag{5}$$

For any considered basis I , and $\forall k \in N$, we have:

$$\nabla f(x^k).y^k = d_k^{\bar{I}}.y_{\bar{I}}^k \quad (\text{since } Ay^k = 0) \tag{6}$$

and from the finite number of possible bases, $\exists\, \beta > 0$ such that:

$$|y^k|^2 \leq \beta^2 |y_{\bar{I}}^k|^2 \tag{7}$$

The move from x^k to z^k is made in a finite number p_k of intermediate steps (kh), $h = 1,2,\ldots,p_k$, corresponding to the displacements $\theta_{kh}y^{kh}$.

Putting $\theta_k = \sum_{h=1}^{pk} \theta_{kh}$, we have:

$$\theta_k \leq 1, \forall k \in N \tag{8}$$

where the strict inequality may hold if a vanishing basic variable breaks the intermediate steps: order 3.1

of the algorithm, with $\theta_k = 0$, $z^k = x^k$, and a pivotal operation, or order 3.2, with $0 < \theta_k < 1$, $z^k \neq x^k$, the maximization of f on $[x^k, z^k]$, and possibly a pivotal operation.

Setting $C_k = \{x_{\bar{I}k} | x_{\bar{I}k} \geq 0\}$, we note that the intermediate steps (kh) are related to the projections of the different parts of the segment $[x^k_{\bar{I}k}, x^k_{\bar{I}k} + \theta_k d_k^{\bar{I}k}]$ on C_k. Therefore:

$$\left.\begin{array}{l} [x^k, z^k]_{\bar{I}k} \subset C_k \\ \\ z^k_{\bar{I}k} = \mathrm{proj}_{C_k}(x^k_{\bar{I}k} + \theta_k d_k^{\bar{I}k}) \end{array}\right\} \Rightarrow z^k_{\bar{I}k} = \ldots$$

$$\ldots \mathrm{proj}_{[x^k, z^k]_{\bar{I}k}}(x^k_{\bar{I}k} + \theta_k d_k^{\bar{I}k}) \ldots$$

$$\ldots \Rightarrow \theta_k d_k^{\bar{I}k} \cdot y^k_{\bar{I}k} \geq |y^k_{\bar{I}k}|^2 \qquad (9)$$

Then, from (5), (6), (7), (8) and (9), and setting $\mu = \min\{1/2\ \beta^2\beta',\ 1/2\}$ we obtain in succession:

$$\mu |y^k_{\bar{I}k}|^2 \le f(x^{k+1}) - f(x^k) \qquad (10)$$

$$\mu \sum_{k=0}^{\infty} |y^k_{\bar{I}k}|^2 \le f(\hat{x}) - f(x^o) < +\infty \text{ from } (H_1)$$

$$\frac{\alpha}{\beta} \sum_{k=0}^{\infty} |y^k| < +\infty$$

$$\sum_{k=0}^{\infty} |x^{k+1} - x^k| < +\infty \text{ because } |x^{k+1} - x^k| \le |y^k|$$

and hence $\sum_{k=0}^{\infty} (x^{k+1} - x^k)$ is a convergent sum. Then:

$$\{x^k \in P \mid k \in N\} \to \overset{*}{x} = x^o + \sum_{k=0}^{\infty} (x^{k+1} - x^k) \in P$$

(ii) <u>The number of pivotal operations cannot be infinite</u>

Let us set $S_* = S(\overset{*}{x})$. There exists an integer k' such that:

$$x^k_{S_*} > 0, \text{ and then } S_* \subset S_k, \ \forall k \ge k' \qquad (11)$$

$$x^k_r = 0 \Longrightarrow r \notin S_* \qquad (12)$$

At each pivotal operation, E_k is increased by an element r (order 5.1 of the algorithm), or is reduced to this element (order 5.2). In the case of an infinite number of pivotings, since E_k cannot increase indefinitely, the second possibility (reduction) will arise an infinite number of times.

After the step k'', corresponding to the first reduction of E_k effective after k', we have from (12):

$$E_k \cap S_* = \emptyset, \ \forall k \geq k'' \tag{13}$$

If after k'', E_k is again reduced another time at a step k, we have by definition (order 5.2 of the algorithm):

$$\nexists s \in (\bar{I}_k - E_k) \cap S_k : T_r^s \neq 0 \tag{14}$$

On the other hand, from (H_2) we can use lemma 3.3 for the point $\overset{*}{x}$ and the basis I_k, which gives:

$$\left.\begin{array}{r} \left.\begin{array}{r} S_* \cap \bar{I}_k \neq \emptyset \\ (13) \end{array}\right\} \Rightarrow S_* \cap (\bar{I}_k - E_k) \neq \emptyset \\ \text{Lemma } 3.3 \end{array}\right\} \Rightarrow$$

$$T_r^{S_*} \cap (\bar{I}_k - E_k) \neq 0 \tag{15}$$

Finally, $\forall k \geq k''$:

$$\left.\begin{array}{r} (11) + (13) + (14) + (15) \Rightarrow S_* \subset I_k \\ (H_2) \end{array}\right\} \Rightarrow$$

$$S_* = I_k$$

Hence, from (12), it is not possible to have a change of basis: a contradiction

5.3. If f is linear, and irrespective of the finite or infinite number of iterations, we have:

$$\exists \alpha > 0 : |y^{kl}_{\bar{I}k}| \geq \alpha \ , \ \forall k$$

for the number of possible bases is finite and $d^{\bar{I}}$ does not depend on x : there is consequently a finite number of possible non-zero values for $y_{\bar{I}}$. The proofs of 5.2.2. can therefore apply and hence we conclude that the number of pivotal operations is finite. The problem Q(I), the equivalent of (P), is a linear programme and, consequently, solving it by the Gradient Projection method ends in a finite number of iterations.

Remark: The same results of convergence are obtained without using the set E_k (i.e. putting down $E_k = \emptyset, \forall k$) but by choosing the candidate s in accordance with the following criterion:

$$s \in \bar{I}_k = x^k_s = \max \{x^k_j \mid j \in \bar{I}_k,\ T^s_r \neq 0\} \qquad (16)$$

(this criterion was originally proposed by Ph. WOLFE [4] who had established finite convergence in the linear case by a proof such as the one that follows).

Indeed, only part 5.2.2.2. (ii) of the proof is to be modified. We can write in its stead,

since $\overset{k}{x} \to \overset{*}{x}$:

$$\exists k' : 0 \le \overset{k}{x}_j < \overset{k}{x}_{j'} , \quad ,\forall j \notin S_*, \forall j' \in S_*, \forall k \ge k' \qquad (17)$$

Furthermore, by applying the lemma 3.3 to the point $\overset{*}{x}$ and the basis I_k:

$$\left.\begin{array}{l} \overset{k}{x}_r = 0 \\ (H_2) \end{array}\right\} \Rightarrow \left.\begin{array}{l} S_* \cap \bar{I}_k \ne \emptyset \\ \text{Lemma 3.3} \end{array}\right\} \Rightarrow T_r^{\bar{I}_k \cap S_*} \ne 0 \qquad (18)$$

Account being taken of the rule of choice (16), we have:

$$(17 + (18) \Rightarrow s \in \bar{I}_k \cap S_* \;, \forall k \ge k'$$

and consequently any variable s entering the basis will no longer be able to leave it. Hence there is contradiction with the hypothesis of an infinity of pivotings.

6. Remarks on the Hypothesis H_2

The hypothesis (H_2) of "non-degeneracy" of the solutions of the system Ax = a is a "natural" hypothesis at the theoretical level, since when we have a given matrix A, only "particular" values of a do not satisfy (H_2). But at the practical level, most problems of real origin have this particularity and do not satisfy (H_2).

The classical theory of disturbances of the second member permits however replacing the original problem (P) by (P_ε), one numerically close to it, satisfying (H_2), by means of the following theorem:

Theorem

A matrix A, of rank m, with rows indexed by an ordered set L , e.g. $L = \{1,2,\dots,m\}$. We consider the function $e:R \to R^m$ defined by:

$$e_i(\varepsilon) = \varepsilon^i, \forall i \in L, \forall \varepsilon \in R$$

Under these conditions, $\forall a \in R^m$, $\exists \alpha > 0$ such that:

$$Ax = a + e(\varepsilon) \Rightarrow \text{rank } A^{S(x)} = m, \forall \varepsilon \in]0,\alpha]$$

where $S(x)$ is the set of indices of the non-zero components of x .

This classical result is based essentially on the fact that the curve describe by $e(\varepsilon)$ can have no part (connected and not reduced to a point) contained in a plane or a linear manifold of R^m .

The following problem (P_ε) can therefore, as a rule, be solved by the Reduced Gradient method:

$$\begin{aligned} \text{Maximize } & f(x) \\ & Ax = a + e(\varepsilon) \\ & x \geq 0 \end{aligned}$$

One can take as small a disturbance $e(\varepsilon)$ as one wishes, but not null, and then it is certain that (P_ε) satisfies (H_2) .

In actual practice, when a is not known, we cannot be certain of choosing ε in the interval $]0,\alpha]$. If ε is taken below the precision of the computer, the same will apply to the disturbances ε^i, $i = 1,2,\ldots,m$, and (P_ε) will practically no longer be distinguishable from (P). This question has to be associated with the study of round-off errors and their influence on a calculation process.

It would be interesting to be able to break away from the effective solving of an approximate problem (P_ε), by adapting the lexographical rule of pivoting, used in the simplex method, to avoid cycling. But, a priori, such an adaptation involves certain difficulties. In fact, while in the simplex method the basic variables can take only a finite number of possible values, this is no longer true with the Reduced Gradient method. In particular, when $x^k \to x^*$, some components of x^k can tend towards zero. The classical theory on which the lexographical rule is based is no longer valid, since it disregards any disturbance value in the presence of any non-null basic variable.

7. References

[1] Faure, P. et Huard, P. (1965) "Résolution de programmes mathématiques à fonction non linéaire par la méthode du gradient réduit". Rev. Fr. R.O. (36), pp. 167-206.

[2] Goldstein, A. A. (1967) "Constructive Real Analysis" Harper and Row (cf. pp. 125-128).

[3] Rosen, J. B. (1960) "The Gradient Projection method for nonlinear programming Part I - Linear constraints" - S.I.A.M. Journal 8 (1) pp. 181-217.

[4] Wolfe, Ph.
1°) (June 22, 1962) "The Reduced Gradient Method" - Rand document.
2°) (1963) "Methods of Nonlinear Programming" - In Recent Advances in Mathematical Programming (Graves-Wolfe eds.) McGraw-Hill pp. 67-86.

[5] Wolfe, Ph., "On the convergence of Gradient Methods under Constraints"
1°) (March 1, 1966) IBM Res. Report RZ-204.
2°) (1972) IBM Journal of Res. and Dev. 16 pp. 407-411.

A QUASI-NEWTON METHOD FOR UNCONSTRAINED MINIMIZATION PROBLEMS

by

Klaus Ritter

ABSTRACT

A method is described for the minimization of a function $F(x)$ of n variables. Convergence to a stationary point is shown without assumptions on second order derivatives. If the sequence generated by this method has a cluster point in a neighbourhood of which $F(x)$ is twice continuously differentiable and has a positive definite Hessian matrix, then the convergence is superlinear. It is shown that under appropriate assumptions n consecutive search directions are conjugate. No computation of second order derivatives is required.

1. Introduction

The basic idea of a quasi-Newton method for the minimization of a function $F(x)$ is to approximate the Hessian matrix of $F(x)$ by a matrix which is updated at each iteration. The difference between known quasi-Newton methods is based on different updating policies and different step size selections. In general the updating schemes use differences of gradients and the step size procedures aim at step size one or at the optimal

---

Sponsored by the United States Army under contract No. DA-31-124-ARO-D-462.

step size or an approximation of it.

In [6] Goldstein and Price describe a method in which a complete new approximation matrix is determined at each iteration. If F(x) depends on n variables, the computation of n values of the gradient of F(x) is required. The step size used is one after a finite number of iterations. The method has the advantage that the properties of the approximating matrix and the rate of the super-linear convergence can easily be established. In the variable metric method [2], [4] the optimal step size is used and the updating procedure requires no additional evaluation of gradients. However, even though Powell [8] proved that the variable metric method is a superlinearly convergent method, the rate of superlinear convergence has not yet been established.

In [9] a quasi-Newton method is described in which the difference of the two most recent values of the gradient is used in the updating procedure, provided a certain test is satisfied. This test is used in order to guarantee convergence without assumptions on second order derivatives of F(x). If the test is not satisfied, an additional gradient evaluation is required. The purpose of this paper is to show that the frequency of the additional gradient evaluation depends on the step size procedure and to present a simple modification of the algorithm for which, under appropriate assumptions, no additional gradient evaluation is required.

2. Some basic properties of quasi-Newton methods

Let $x \in E^n$ and assume that $F(x)$ is a real valued function. If $F(x)$ is differentiable at a point x_i, we denote its gradient at x_i by $F(x_i)$ or g_i. If $F(x)$ is twice differentiable at x_i, we denote the Hessian matrix of $F(x)$ at x_i by $G(x_i)$ or G_i. A prime is used to indicate the transpose of a vector or a matrix. For any $x \in E^n$, $||x||$ denotes the Eucledian norm of x and, for any (n,n)-matrix M, $||M||$ denotes the matrix norm induced by the Euclidean vector norm.

In order to simplify the following discussion and the description of the algorithm we shall assume throughout sections 2 - 5 that $F(x)$ satifies the following

Assumption 1

$F(x)$ is twice continuously differentiable and there are real numbers $0 < \mu < \eta$ such that

$$\mu||x||^2 \leqq x'G(y)x \leqq \eta||x||^2 \quad \text{for all } x,y \in E^n.$$

Since it is in general difficult to verify Assumption 1 for a given $F(x)$ we shall describe a modified algorithm in section 6 for which super-linear convergence can be established under much weaker assumptions.

A point x is said to be a stationary point of $F(x)$ if $\nabla F(x) = 0$. It is well-known that Assumption 1 implies that $F(x)$ has exactly one stationary point, say z, and that $F(z) < F(x)$ for every $x \neq z$. It is the purpose of this paper to describe a new quasi-Newton method which either terminates after a finite number of steps with z or generates an infinite sequence $\{x_j\}$ which converges superlinearly to z. Here

$$x_{j+1} = x_j - \sigma_j s_j ,$$

where $s_j \in E^n$ is the search direction and $\sigma_j \in E^1$ is an appropriate step size.

In [7], [10] an algorithm for the minimization of F(x) which generates a sequence $\{x_j\}$ with $x_{j+1} = x_j - \sigma_j s_j$ is said to be a quasi-Newton method if $\{x_j\}$ converges to some z such that $\nabla F(z) = 0$ and

$$(1) \qquad \frac{1}{||g_j||} \, ||\sigma_j G(z) s_j - g_j|| \to 0 \text{ as } j \to \infty .$$

It is shown that then $\{x_j\}$ converges superlinearly to z, i.e.,

$$\frac{||x_{j+1} - z||}{||x_j - z||} \to 0 \quad \text{as } j \to \infty .$$

In order to obtain a sequence $\{\sigma_j s_j\}$ for which (1) is satisfied it is often convenient to set

$$(2) \qquad s_j = M_j g_j$$

where $\{M_j\}$ is a sequence of (n,n)-matrices such that

$$(3) \qquad ||M_j - G^{-1}|| \to 0 \quad \text{as } j \to \infty , \; G = G(z)$$

and to determine σ_j such that

$$(4) \qquad \sigma_j \to 1 \text{ as } j \to \infty$$

or

$$\sigma_j = 1 \quad \text{for j sufficiently large.}$$

It is well-known that, even with $M_j = G_j^{-1}$ for all j, it may not be possible to use the step size 1 in the early stages of the iteration, if $\{x_j\}$ is to converge to z.

In a practical algorithm we shall in general choose $M_j \neq G_j^{-1}$ in order to avoid the computation of second order partial derivatives of F(x). Instead we try to determine M_j by using appropriate differences of gradients of F(x). A typical example for such an approach is the Goldstein-Price method [6]. In this method the determination of M_j requires the evaluation of the gradient of F(x) at n appropriately chosen points. It is shown in [6] that after a finite number of iterations the step size 1 is acceptable.

Since in this method no attempt is made to use values of $\nabla F(x)$ at previously encountered points in the construction of M_j, it appears to be reasonable to choose $\sigma_j = 1$ as soon as possible. If on the other hand we intend to keep the number of additional gradient evaluations at a minimum by trying to use $g_j, g_{j-1}, \ldots, g_{j-n}$ in the construction of M_j, a different choice of σ_j could be appropriate.

In the next section we shall briefly describe a method, given in [9], which attempts to update M_j by using g_j and g_{j+1} whenever possible. It will turn out that in this method the number of additional gradient evaluations depends decisively on the choice of σ_j.

3. The influence of the step size

The basic idea of the method given in [9] is to try to use $g_j, \ldots, g_{j-n}$ in the construction of M_j. Ideally, such an approach would not require any additional evaluation of the gradient of $\nabla F(x)$. Unfortunately, it does not always seem to be possible to determine M_j from $g_j, \ldots, g_{j-n}$ if

$$(1) \qquad ||M_j - G^{-1}|| \to 0 \quad \text{as } j \to \infty,\ G = G(z)$$

is to hold. In order to see this we shall try to define M_j in terms of $g_j, \ldots, g_{j-n}$ by using the method suggested in [9].

Let

$$(2) \qquad d_{ij} = \frac{||g_{j-i} - g_{j-i+1}||}{||\sigma_{j-i} s_{j-i}||}, \quad i = 1, \ldots, n$$

$$(3) \qquad p_{ij} = \frac{s_{j-i}}{||s_{j-i}||}, \qquad i = 1, \ldots, n.$$

Using Taylor's theorem we obtain

$$(4) \qquad d_{ij} = Gp_{ij} + w_{ij}, \qquad i = 1, \ldots, n$$

where $G = G(z)$, $w_{ij} \in E^n$ and, for $i = 1, \ldots, n$,

$$(5) \qquad ||w_{ij}|| \to 0 \quad \text{as } j \to \infty \quad \text{and } x_j \to z .$$

Now we define the matrices D_j, P_j, and W_j by

$$D_j' = (d_{1j}, \ldots, d_{nj}), \quad P_j' = (p_{1j}, \ldots, p_{nj})$$

and

$$W_j' = (w_{1j}, \ldots, w_{nj}) .$$

Then it follows from (2) - (4) that

$$(6) \qquad D_j = P_j G + W_j \quad .$$

Assuming that D_j is nonsingular and multiplying (6) from the left by D_j^{-1} and from the right by G^{-1} we obtain

$$(7) \qquad G^{-1} = D_j^{-1} P_j + D_j^{-1} W_j G^{-1} .$$

Let $M_j = D_j^{-1} P_j$. If we assume that D_j^{-1} exists for all j and that $\{D_j^{-1}\}$ is bounded, then it follows from (5) and (7) that

$$||D_j^{-1} W_j G^{-1}|| \to 0 \text{ if } x_j \to z$$

and, therefore,

$$||M_j - G^{-1}|| \to 0 \quad \text{if} \quad x_j \to z \ .$$

Thus $M_j = D_j^{-1} P_j$ satisfies (1) if $\{D_j^{-1}\}$ exists and is bounded and $x_j \to z$ as $j \to \infty$. In order to find the conditions under which $\{D_j^{-1}\}$ is bounded we observe that, similar to the definition of D_j', we have

$$D_{j+1}' = (d_{1,j+1}, \ \ldots \ ,d_{n,j+1})$$

where
$$d_{1,j+1} = \frac{g_j - g_{j+1}}{||\sigma_j s_j||} =: d_j \ , \ d_{i,j+1} = d_{i-1,j} \ ,$$
$$i = 2, \ \ldots \ ,n$$

and $d_{ij}, \ \ldots \ ,d_{n-1,j}$ are defined by (2).

Set $T_j = \text{span}\{d_{ij}, \ \ldots \ ,d_{n-1,j}\}$ and let $d_j = u_j + v_j$ where $u_j \in T_j$ and v_j is orthogonal to T_j. Since we assumed the existence of D_j^{-1}, the vectors $d_{ij}, \ldots, d_{n-1,j}$ are linearly independent. Thus D_{j+1}^{-1} exists if and only if $v_j \neq 0$. Furthermore, a necessary condition for $\{D_j^{-1}\}$ to be bounded is that

(8) $\quad ||v_j|| \geqq \varepsilon$ for some $\varepsilon > 0$ and all j.

If we set $D_j^{-1} = (c_{1j}, \ldots, c_{nj})$, then c_{nj} is orthogonal to T_j and

$$|c'_{nj} d_j| = |c'_{nj} v_j| = ||c_{nj}|| \; ||v_j|| .$$

Thus (8) is satisfied if and only if

(9) $$|c'_{nj} d_j| \geqq \varepsilon ||c_{nj}|| .$$

With $s_j = D_j^{-1} P_j g_j$ it follows from Lagrange's formula [11] that, for every choice of $\sigma_j > 0$,

(10) $$|c'_{nj} d_j| = \frac{|c'_{nj}(g_j - g_{j+1})|}{||\sigma_j s_j||} = |c'_{nj} G(\xi_j) \frac{s_j}{||s_j||}|$$

where $\xi_j \in \{x | x = x_j - t\sigma_j s_j, \; 0 \leqq t \leqq 1\}$. From (10) it is clear that (9) need not always be satisfied.

In [9] the inequality (9) is used to test whether d_j is acceptable as the first column of D'_{j+1}. If (9) is not satisfied d_j is rejected and replaced by

$$\bar{d}_j = \frac{g_j - \nabla F(x_j - ||\sigma_j s_j|| \frac{c_{nj}}{||c_{nj}||})}{||\sigma_j s_j||} .$$

Thus an additional gradient evaluation is necessary whenever (9) is not satisfied.

The matrices D'_{j+1} and P'_{j+1} are defined as follows:

$$D'_{j+1} = (d_{1,j+1}, \dots, d_{n,j+1}) ,$$
$$P'_{j+1} = (p_{1,j+1}, \dots, p_{n,j+1})$$

where

$$d_{i,j+1} = d_{i-1,j},\ p_{i,j+1} = p_{i-1,j},\ i = 2, \dots, n$$

$$d_{1,j+1} = \begin{cases} d_j & \text{if (9) is satisfied} \\ \bar{d}_j & \text{if (9) is not satisfied} \end{cases}$$

$$p_{1,j+1} = \begin{cases} \dfrac{s_j}{||s_j||} & \text{if (9) is satisfied} \\ \dfrac{c_{nj}}{||c_{nj}||} & \text{if (9) is not satisfied.} \end{cases}$$

In [9] it is shown that then the sequences $\{D_j\}$, $\{D_j^{-1}\}$ are bounded and that

$$||D_j^{-1} P_j - G^{-1}|| \to 0 \quad \text{as} \quad j \to \infty .$$

Furthermore, for j sufficiently large either $\sigma_j = 1$ or σ_j equal to an appropriate approximation of the optimal step size is acceptable.

In the following we shall assume these results and investigate the relationship between the choice of σ_j and the frequency with which (9) is satisfied. It will turn out that, if we wish to satisfy (9) as often as possible, it is advisable to use a step size procedure which approximates the optimal step size rather than to aim at $\sigma_j = 1$.

Assuming $\varepsilon < \mu$ we conclude from (10) that (9)

will be satisfied for j sufficiently large, if

$$(11) \qquad \left|\left|\frac{s_j}{||s_j||} \pm \frac{c_{nj}}{||c_{nj}||}\right|\right| \to 0 \quad \text{as} \quad j \to \infty.$$

Since the sequences $\{D_j^{-1}\}$and $\{||g_j|| \; ||s_j||^{-1}\}$are bounded,

$$s_j = D_j^{-1}P_jg_j = \sum_{i=1}^{n} c_{ij}p_{ij}'g_j$$

implies that (11) is satisfied if and only if

$$(12) \qquad \frac{|p_{ij}'g_j|}{||g_j||} \to 0 \quad \text{as} \quad j \to \infty \text{ , } i = 1, \ldots ,n-1 \text{ .}$$

In order to derive sufficient conditions for (12) to hold we assume throughout the remainder of this section that G(x) satisfies a Lipschitz condition, i.e., there is a constant L such that

$$||G(x) - G(y)|| \leqq L||x - y|| \quad \text{for all } x,y \in E^n.$$

Suppose first that we try to use $\sigma_j = 1$. Let j be so large that $\sigma_j = 1$ is acceptable. Then it follows from Lagrange's formula that for every $i \in \{1, \ldots ,n\}$ there is $\xi_{ij} \in \{x|x = x_j - ts_j, \; 0 \leqq t \leqq 1\}$ such that

$$\begin{aligned} p_{ij}'g_{j+1} &= p_{ij}'g_j - p_{ij}'G(\xi_{ij})s_j \\ &= p_{ij}'g_j - p_{ij}'GD_j^{-1}P_jg_j \\ &\qquad - p_{ij}'(G(\xi_{ij}) - G)D_j^{-1}P_jg_j \\ &= (d_{ij}' - p_{ij}'G)D_j^{-1}P_jg_j \\ &\qquad - p_{ij}'(G(\xi_{ij}) - G)D_j^{-1}P_jg_j \quad , \end{aligned}$$

and

$$(13)\quad |p'_{ij}g_{j+1}| \leq (||d_{ij} - Gp_{ij}|| + ||G(\xi_{ij}) - G||) \times ||D_j^{-1}p_j|| \; ||g_j|| \; .$$

From (2), (3) and Taylor's theorem [3] it follows that

$$(14)\quad ||d_{ij} - Gp_{ij}|| = ||(\int_0^1 G(x_{j-i}-ts_{j-i})dt - G)p_{ij}|| \leq \sup_{0\leq t\leq 1} ||G(x_{j-i}-ts_{j-i}) - G|| \leq L||x_{j-i}-x_{j-i+1}||+L||x_{j-i+1}-z||.$$

Thus we conclude from (13), (14) and Lemmas 2 and 4 that

$$|p'_{ij}g_{j+1}| = 0(||g_{j-i}|| \; ||g_j||) \; , \; i = 1, \; ... \; ,n \; .$$

By the definition of P_{j+1} this implies

$$\frac{|p'_{i,j+1}g_{j+1}|}{||g_{j+1}||} = \begin{cases} \dfrac{|p'_j g_{j+1}|}{||g_{j+1}||} & \text{if } d_{1,j+1} = d_j \\ \dfrac{|c'_{nj}g_{j+1}|}{||g_{j+1}||} & \text{if } d_{1,j+1} = \bar{d}_j \end{cases}$$

$$(15)\quad \frac{|p'_{i+1,j+1}g_{j+1}|}{||g_{j+1}||} = 0(\frac{||g_j|| \; ||g_{j-i}||}{||g_{j+1}||}) \; ,$$

$$i = 1, \; ... \; ,n-1.$$

By Theorem 3 in [9], $||g_{j+1}|| = 0(||g_j|| \; ||g_{j-n}||)$. Let $||g_{j+1}|| = \delta_j||g_j|| \; ||g_{j-n}||$ and suppose $\delta_j \geq \delta > 0$ for all j. Then (15) implies

$$(16)\quad \frac{|p'_{i+1,j+1}g_{j+1}|}{||g_{j+1}||} = 0(\frac{||g_{j-1}||}{||g_{j-n}||}) \to 0 \text{ as } j \to \infty,$$

$$i = 1,...,n-1 \; ,$$

and it follows from Lemma 2 that

$$(17)\quad \frac{|p'_{1,j+1}g_{j+1}|}{\|g_{j+1}\|} \geqq \rho_1 > 0 \text{ for } j \text{ sufficiently large.}$$

Therefore, (12) does not hold and we cannot guarantee that, for j sufficiently large, (9) is satisfied, i.e., d_j is acceptable for the updating of D_j^{-1}. On the contrary we can now show that, if $\delta_j \geqq \delta > 0$ for all j, then there is $j(\varepsilon)$ such that $d_{1,j(\varepsilon)+1} = \bar{d}_{j(\varepsilon)}$ implies $d_{1,j+1} = \bar{d}_j$ for all $j \geqq j(\varepsilon)$, i.e., if d_j is rejected as first column of D'_{j+1} for some sufficiently large j, all subsequent d_j's will be rejected.

Indeed, (16) and (17) imply that, if $\delta_j \geqq \delta > 0$ for all j, then

$$(18)\quad \left\| \frac{s_{j+1}}{\|s_{j+1}\|} \pm \frac{c_{1,j+1}}{\|c_{1,j+1}\|} \right\| \to 0 \text{ as } j \to \infty .$$

By Taylor's theorem,

$$(19)\quad \bar{d}_j = G\frac{c_{nj}}{\|c_{nj}\|} + \left(\int_0^1 G(x_j - t\|s_j\|\frac{c_{nj}}{\|c_{nj}\|})dt - G\right)\frac{c_{nj}}{\|c_{nj}\|}$$

$$(20)\quad d_{j+1} = G\frac{c_{1,j+1}}{\|c_{1,j+1}\|} + G\left(\frac{s_{j+1}}{\|s_{j+1}\|} - \frac{c_{1,j+1}}{\|c_{1,j+1}\|}\right) + \left(\int_0^1 G(x_{j+1} - ts_{j+1})dt - G\right)\frac{s_{j+1}}{\|s_{j+1}\|} .$$

Since by Lemma 1 in [9], $c_{1,j+1}$ is a multiple of c_{nj}, (18)- (20) imply

$$(21)\quad \|d_{j+1} \pm \bar{d}_j\| \to 0 \text{ as } j \to \infty .$$

If $d_{1,j+1} = \bar{d}_j$, then $\bar{d}'_j c_{n,j+1} = 0$ and it follows from (21) that

$$|d'_{j+1}c_{n,j+1}| < \varepsilon||c_{n,j+1}|| \text{ for sufficiently large } j .$$

If F(x) is an arbitrary function that satisfies Assumption 1, there is in general no reason to expect that the actual rate of convergence is faster than the predicted rate of convergence. Thus the above result implies that in general d_j will very often not be acceptable for the updating of D_j if $\sigma_j = 1$ is used.

Next let us assume that instead of $\sigma_j = 1$ we use a σ_j which is an approximation to the optimal step size with the property

$$(22) \quad |g'_{j+1}s_j| = O(||g_j||^2||s_j||) .$$

A simple method to determine such a σ_j is described in the next section.

Let j be so large that this choice of σ_j is acceptable. Then it follows again from Lagrange's formula that, for i = 1, ... ,n ,

$$(23) \quad \begin{aligned} p'_{ij}g_{j+1} &= p'_{ij}g_j - \sigma_j p'_{ij}G(\xi_{ij})s_j \\ &= (1-\sigma_j)p'_{ij}g_j+\sigma_j(d'_{ij}-p'_{ij}G)D_j^{-1}P_jg_j - \\ &\quad - \sigma_j p'_{ij}(G(\xi_{ij}) - G)D_j^{-1}P_jg_j , \end{aligned}$$

where $\xi_{ij} \in \{x|x = x_j - t\sigma_j s_j, 0 \leqq t \leqq 1\}$. Using (14), (23) and Lemma 4 we obtain, for i=1,...,n ,

$$(24) \quad |p'_{ij}g_{j+1}| = |1-\sigma_j||p'_{ij}g_j| + O(||g_{j-i}|| \, ||g_j||),$$

Now suppose $d_{1,j+1} = d_j$ and

$$(25) \qquad \frac{|p_{ij}' g_j|}{||g_j||} = O\left(\frac{||g_{j-i}||}{||g_{j-n}||}\right), \quad i = 1, \ldots, n-1 .$$

Since by Lemma 6, $|1-\sigma_j| = O(||g_{j-n}||)$ it follows then from (22), (24), and the definition of P_{j+1} that

$$\frac{|p_{1,j+1} g_{j+1}|}{||g_j||} = O(||g_j||)$$

$$\frac{|p_{i+1,j+1}' g_{j+1}|}{||g_j||} = O(||g_{j-i}||), \quad i = 1, \ldots, n-1.$$

Observing that $||g_{j-i}|| = O(||g_{j-n+1}||)$ for $i = 0, \ldots, n-1$ and using Lemma 2 we obtain

$$(26) \qquad ||g_{j+1}|| = O(||g_j|| \; ||g_{j-n+1}||) .$$

In order to see under which conditions (25) is satisfied set $||g_{j+1}|| = \delta_j ||g_j|| \; ||g_{j-n+1}||$. Suppose $\delta_j \geqq \delta > 0$ for all j, i.e., the actual rate of convergence is not faster than the rate predicted by (26). Replacing j by j-1 and assuming $d_{1j} = d_{j-1}$ we conclude from (22) that

$$(27) \qquad \frac{|p_{1j}' g_j|}{||g_j||} = \frac{|s_{j-1}' g_j|}{||s_{j-1}|| \; ||g_j||} = O\left(\frac{||g_{j-1}||^2}{\delta_{j-1} ||g_{j-1}|| \cdot ||g_{j-n}||}\right)$$

$$= O\left(\frac{||g_{j-1}||}{||g_{j-n}||}\right),$$

i.e., (25) holds for $i = 1$. Now suppose that $d_{1,j+1} = d_j$ and (25) is true for some $i < n-1$. Since $p_{i+1,j+1} = p_{ij}$, it follows from (24) and $|1 - \sigma_j| = O(||g_{j-n}||)$ that

$$(28) \quad \frac{|p_{i+1,j+1}' g_{j+1}|}{||g_{j+1}||} = 0\left(\frac{||g_{j-n}||}{\delta_j ||g_{j-n+1}||} \frac{||g_{j-i}||}{||g_{j-n}||} + \frac{||g_{j-i}||}{\delta_j ||g_{1-n+1}||}\right)$$

$$= 0\left(\frac{||g_{j-i}||}{||g_{j-n+1}||}\right) ;$$

i.e., (25) holds for j+1 and i+1.

It follows now from (27) and (28) that (25) holds for all j sufficiently large for which the following two assumptions are satisfied:

$$(29) \qquad \delta_{j-i} \geqq \delta > 0 \ , \ i = 1, \ \ldots \ ,n$$

$$(30) \qquad d_{1,j+1-i} = d_{j-i} \ , \ i = 1, \ \ldots \ ,n \ .$$

Because (25) implies (12), we have the following result: If $\delta_j \geqq \delta > 0$ for all j and if d_j is accepted as the first column of D_{j+1}' for n consecutive sufficiently large j, then all subsequent d_j' s will be accepted.

In proving the above result we used (30) to guarantee that, because of $p_{1,j+1} = s_j/||s_j||$ and (22),

$$(31) \qquad |p_{1,j+1}' g_{j+1}| = 0(||g_j||^2).$$

Using the algorithm given in [9] with a step size procedure for which (22) holds it cannot be shown that (30) is satisfied. Therefore, we need a modified algorithm for which (31) is always satisfied. Such an algorithm is given in the next section.

4. The Algorithm

Throughout this section we assume that Assumption 1 is satisfied. The algorithm starts with an arbitrary $x_0 \in E^n$, an arbitrary (n,n)-matrix P_0 and an arbitrary nonsingular D_0^{-1}.

In the j-th cycle of the algorithm the matrix P_{j+1} is computed from P_j by deleting the last column of P_j, shifting all other columns one place to the right and inserting an appropriate first column. D_{j+1} is obtained in a similar way. In order to have a convenient expression for this procedure we introduce the following notation: For any (n,n)-matrix $P' = (p_1, \ldots, p_n)$ and any $a \in E^n$ with $a \neq 0$, let

$$\Psi(P',a) = (a/\|a\|, p_1, p_2, \ldots, p_{n-1}) .$$

For any (n,n)-matrix $D^{-1} = (c_1, \ldots, c_n)$ and any $a \in E^n$ with $c_n'a \neq 0$, let

$$\Phi(D^{-1},a) = \left(\frac{c_n}{c_n'a}, c_1 - \frac{c_1'a}{c_n'a}c_n, \ldots, c_{n-1} - \frac{c_{n-1}'a}{c_n'a}c_n\right) .$$

Denoting the columns of D' by $d_1, \ldots, d_n$, it is easy to verify that with $\tilde{D}' = (a, d_1, \ldots, d_{n-1})$, we have $\tilde{D} = (\Phi(D^{-1},a))^{-1}$.

We give now a detailed description of the algorithm followed by a general discussion which gives a motivation for the various steps. At the beginning of the j-th cycle the following data is available: $x_j \in E^n$, $g_j = \nabla F(x_j)$, $D_j^{-1} = (c_{1j}, \ldots, c_{nj})$ and $P_j' = (p_{1j}, \ldots, p_{nj})$. The j-th cycle consists of the following four steps:

Algorithm 1

Step I: Determination of s_j

Set

$$\bar{s}_j = D_j^{-1} P_j g_j$$

and

$$s_j = \begin{cases} \bar{s}_j & \text{if } g_j \bar{s}_j \geqq \alpha_j ||g_j||^2 \\ -\bar{s}_j & \text{if } -g_j \bar{s}_j \geqq \alpha_j ||g_j||^2 \\ g_j & \text{if } |g_j \bar{s}_j| < \alpha_j ||g_j||^2 \end{cases},$$

where $\alpha_j = \min \{\alpha, ||g_j||\}$, $0 < \alpha < 1$. Go to Step II.

Step II: Determination of σ_j

Compute $F(x_j - s_j)$ and $\omega_j = F(x_j - s_j) - F(x_j) + g_j s_j$. Set

$$\bar{\sigma}_j = \frac{g_j s_j}{2\omega_j}$$

and let ν_j be the smallest nonnegative integer such that

$$F(x_j) - F(x_j - 0.5^{\nu_j} \bar{\sigma}_j s_j) \geqq \gamma_1 0.5^{\nu_j} \bar{\sigma}_j g_j s_j, \quad 0 < \gamma_1 < \tfrac{1}{2}.$$

Set

$$\sigma_j = (0.5)^{\nu_j} \bar{\sigma}_j, \quad \tilde{x}_{j+1} = x_j - \sigma_j s_j$$

and compute $\tilde{g}_{j+1}$. If $\tilde{g}_{j+1} = 0$ stop; otherwise go to Step III.

Step III: Updating of D_j^{-1} and P_j

Let

$$d_j = \frac{g_j - \tilde{g}_{j+1}}{||\sigma_j s_j||} \quad \text{and} \quad p_j = \frac{s_j}{||s_j||}.$$

If
$$|d_j' c_{nj}| \geqq \gamma_2 ||c_{nj}|| \ , \ \gamma_2 > 0$$
set
$$D_{j+1}^{-1} = \Phi(D_j^{-1}, d_j), \ P_{j+1} = \Psi(P_j, p_j) \ ,$$
$$x_{j+1} = \tilde{x}_{j+1} \quad \text{and} \quad g_{j+1} = \tilde{g}_{j+1} \ .$$
Replace j by j+1 and go to Step I.

If
$$|d_j' c_{nj}| < \gamma_2 ||c_{nj}||, \text{ go to Step IV.}$$

Step IV: Special Updating Step

Compute
$$\tilde{d}_j = (g_j - \nabla F(x_j - ||\sigma_j s_j|| \frac{c_{nj}}{||c_{nj}||})) \, ||\sigma_j s_j||^{-1}$$
and set
$$D_{j+1}^{-1} = \Phi(D_j^{-1}, \tilde{d}_j) \text{ and } P_{j+1} = \Psi(P_j, \frac{c_{nj}}{||c_{nj}||}) \ .$$
Let
$$\tilde{s}_{j+1} = D_{j+1}^{-1} P_{j+1} \tilde{g}_{j+1} \ .$$
If $\tilde{g}_{j+1}' \tilde{s}_{j+1} \leqq 0$, choose $x_{j+1} = \tilde{x}_{j+1}$ and $g_{j+1} = \tilde{g}_{j+1}$; otherwise let ν_j be the smallest nonnegative integer such that
$$F(\tilde{x}_{j+1} - 0.5^{\nu_j} \tilde{s}_{j+1}) \leqq F(\tilde{x}_{j+1}) \ ,$$
set $x_{j+1} = \tilde{x}_{j+1} - 0.5^{\nu_j} \tilde{s}_{j+1}$ and compute g_{j+1}. If $g_{j+1} = 0$, stop; otherwise replace j by j+1 and go to Step I.

By Lemma 2, $||D_j^{-1} P_j - G^{-1}|| \to 0$ as $j \to \infty$. Thus $\bar{s}_j = D_j^{-1} P_j g_j$ is a desirable search vector for j sufficiently large. However, in the early stages of the iteration it could happen that $g_j\bar{s}_j \leqq 0$. Thus $\bar{s}_j$ is accepted as search direction only if

$$(1) \qquad g_j\bar{s}_j \geqq \alpha_j||g_j|| \quad .$$

From (1) in the proof of Lemma 3 it follows that $g_j\bar{s}_j \geqq \frac{1}{2\eta}||g_j||^2$ for j sufficiently large. Since $||g_j|| \to 0$, $\alpha_j \to 0$ and (1) is satisfied after a finite number of steps. If η is known it suffices, of course, to choose $\alpha_j = \alpha < \frac{1}{2\eta}$.

Let σ_j^* denote the optimal step size. By Lagrange's formula there is $\xi_j \in \{x|x = x_j - t\sigma_j^* s_j,\ 0 \leqq t \leqq 1$ such that

$$0 = s_j'\nabla F(x_j - \sigma_j^* s_j) = g_j's_j - \sigma_j^* s_j'G(\xi_j)s_j$$

or, for j sufficiently large,

$$\sigma_j^* = \frac{g_j'D_j^{-1}P_jg_j}{(D_j^{-1}P_jg_j)'GD_j^{-1}P_jg_j + s_j'(G(\xi_j) - G)s_j} .$$

Since $D_j^{-1}P_j \to G^{-1}$, $\sigma_j^* \to 1$, and it appears reasonable to use $F(x_j - s_j)$ in addition to $F(x_j)$ and g_js_j in trying to approximate σ_j^*. If $Q_j(\sigma)$ denotes the quadratic function in σ for which

$$Q(0) = F(x_j),\ \frac{d}{d\sigma}Q(0) = -g_js_j,\ Q_j(1) = F(x_j - \sigma_j)$$

holds, it follows immediately that $\omega_j > 0$ and that

$$\bar{\sigma}_j = \frac{g_js_j}{2\omega_j}$$

is the global minimizer of $Q_j(\sigma)$. We choose $\bar{\sigma}_j$ as an approximation to the optimal step size. In order to guarantee convergence $\bar{\sigma}_j$ is accepted as step size only if it satisfies a test of the type introduced by Goldstein [5] and Armijo [1]. Since by Taylor's theorem

$$\lim_{\sigma\to 0} \frac{F(x_j) - F(x_j - \sigma s_j)}{g_j s_j} = 1 ,$$

ν_j is well-defined. By Lemma 3, $\nu_j = 0$ for j sufficiently large.

In accordance with the discussion in section 3 we try to use $d_j = (g_j - \tilde{g}_{j+1})||\sigma_j s_j||^{-1}$ as the first column of D'_{j+1} provided the test

$$(2) \qquad |d_j' c_{nj}| \geq \gamma_2 ||c_{nj}||$$

is satisfied. Otherwise a special updating step, involving an additional evaluation of $\nabla F(x)$, is performed. Since in the proof of Lemma 1 it is shown that $|\tilde{d}_j' c_{nj}| \geq \mu ||c_{nj}||$, D_{j+1}^{-1} is well defined. However, there is no guarantee that

$$|p'_{1,j+1} \tilde{g}_{j+1}| = \frac{|c_{nj} \tilde{g}_{j+1}|}{||c_{nj}||} = O(||g_j||^2) .$$

Since this property is essential in proving that under appropriate assumptions (2) is satisfied for j sufficiently large, a new point $x_{j+1} = \tilde{x}_{j+1} - 0.5^{\nu_j} \tilde{s}_{j+1}$ is computed. In Lemma 3 it is shown that $\nu_j = 0$ for j sufficiently large, and Lemma 5 states that

$$(3) \qquad |p'_{1,j+1} g_{j+1}| = O(||\tilde{g}_{j+1}|| \; ||g_j||) = O(||g_j||^2)$$

provided G(x) satisfies a Lipschitz condition.

Furthermore, (see Theorem 2)

$$(4) \qquad \frac{||g_{j+1}||}{||\tilde{g}_{j+1}||} \to 0 \quad \text{as } j \to \infty .$$

Both the properties (3) and (4) of the special updating step will be used to establish that under appropriate assumptions (2) is satisfied for j sufficiently large.

5. Convergence Results

Throughout this section we assume that Assumption 1 is satisfied. We shall often use the set $J \subset \{0,1,2, \ldots\}$ which is defined in such a way that $j \in J$ if and only if D^{-1}_{j+1} has been determined by Step IV of the algorithm. First we need the following

Lemma 1

The sequences $\{D_j\}$, $\{D_j^{-1}\}$, and $\{s_j\}$ are bounded.

Proof:

By Lagrange's formula there are

$$\xi_j, \eta_j \in \{x | x = x_j - t||\sigma_j s_j|| \frac{c_{nj}}{||c_{nj}||}, \; 0 \leq t \leq 1\}$$

such that

$$\tilde{d}_j'\tilde{d}_j = \tilde{d}_j' G(\xi_j) \frac{c_{nj}}{||c_{nj}||} \leq \eta ||\tilde{d}_j||$$

and

$$c_{nj}'\tilde{d}_j = c_{nj}' G(\eta_j) \frac{c_{nj}}{||c_{nj}||} \geq \mu ||c_{nj}|| \quad .$$

In a similar way it can be shown that

$$||d_j|| = \frac{||g_j - g_{j+1}||}{||\sigma_j s_j||} \leq \eta \quad .$$

Thus we have, for all j, $||d_{1,j+1}|| \leqq \eta$ and

$$|d_j'c_{nj}| \geqq \gamma_2||c_{nj}||,\ j \notin J;\ |\tilde{d}_j'c_{nj}| \geqq \mu||c_{nj}||,\ j \in J.$$

By the definition of D_{j+1} and D_{j+1}^{-1} this implies that $\{D_j\}$ and $\{D_j^{-1}\}$ are bounded. Finally,

$$||\bar{s}_j|| = ||D_j^{-1}P_jg_j|| \leqq ||D_j^{-1}||\ ||g_j|| .$$

Since $\{g_j\}$ is bounded this completes the proof.

Theorem 1:

i) There is a unique z such that $\nabla F(z) = 0$. Moreover, $F(z) < F(x)$ for all $x \in E^n$, $x \neq z$.

ii) The algorithm either terminates after a finite number of cycles with z or generates an infinite sequence $\{x_j\}$ which converges to z.

Proof:

i) This statement is a well-known consequence of Assumption 1.

ii) If the algorithm terminates with x_j, then $\nabla F(x_j) = 0$ and, therefore, $x_j = z$. Suppose the algorithm generates an infinite sequence. Since $\{x|F(x) \leqq F(x_0)\}$ is compact and $F(x_{j+1}) \leqq F(x_j)$ for all j, $\{x_j\}$ has at least one cluster point. In order to complete the proof it suffices to show that $||g_j|| \to 0$ as $j \to \infty$ because this implies that every cluster point of x_j is a stationary point.

Suppose there is $0 < \varepsilon \leqq \alpha$ and an infinite subset $I \subset \{0,1,2, \ldots\}$ such that $||g_j|| \geqq \varepsilon$ for

$j \in I$. Then $g_j' s_j \geqq \varepsilon^3 =: \delta$ for $j \in I$. By Taylor's theorem there is $\eta_j \in \{x | x = x_j - ts_j,\ 0 \leq t \leq 1\}$ such that

$$(1)\quad \omega_j = F(x_j - s_j) - F(x_j) + g_j s_j = \frac{1}{2} s_j' G(\eta_j) s_j \leqq \frac{\eta}{2} ||s_j||^2.$$

Since $\{s_j\}$ is bounded it follows that there is $\tau_1 > 0$ such that

$$(2)\quad \bar{\tau}_j = \frac{g_j s_j}{2\omega_j} \geqq \frac{\delta}{\eta ||s_j||^2} \geqq \tau_1 \quad \text{for } j \in I\ .$$

Furthermore, it follows again from Taylor's theorem that there is $\xi_j \in \{x | x_j - t\sigma s_j,\ 0 \leq t \leq 1\}$ such that

$$F(x_j) - F(x_j - \sigma s_j) = \sigma g_j s_j + (\nabla F(\xi_j) - g_j)' s_j \sigma$$
$$\geqq \sigma g_j s_j - ||\nabla F(\xi_1) - g_j||\ ||s_j||\ \sigma$$

Since $\nabla F(x)$ is uniformly continuous and $\{s_j\}$ is bounded this implies that there is $\tau_2 > 0$ such that

$$(3)\quad \frac{F(x_j) - F(x_j - \sigma s_j)}{\sigma g_j' s_j} \geqq 1 - \frac{||\nabla F(\xi_j) - g_j||\ ||s_j||}{g_j s_j} \geqq \gamma_1$$

for $0 < \sigma \leqq \tau_2$ and $j \in I$.

Using (2), (3) and the definition of ν_j in Step Step II of the algorithm we have $\sigma_j \geqq \tau := \min\{\tau_1, 0.5\tau_2\}$ for $j \in I$. Therefore,

$$F(x_j) - F(\tilde{x}_{j+1}) \geqq \gamma_1 \delta\tau > 0 \quad \text{for } j \in I\ .$$

This implies that $F(x_j) \to -\infty$ as $j \to \infty$. Since $F(x) \geqq F(z)$ for all x, this is a contradiction and it follows that $||g_j|| \to 0$ as $j \to \infty$.

In order to prove that $\{x_j\}$ converges superlinearly to z we need some further properties of the algorithm which are established in the following 3 lemmas.

Lemma 2

Let $G = G(z)$, $P_j' = (p_{1j},\ldots,p_{nj})$ and $D_j' = (d_{1j},\ \ldots\ ,d_{nj})$. Then

i) $p_{ij} = \dfrac{s_{j-i}}{||s_{j-i}||}$, $d_{ij} = Gp_{ij}+E_{j-i}p_{ij}$ if $j-i \notin J$

$p_{ij} = \dfrac{c_{n,j-i}}{||c_{n,j-i}||}$, $d_{ij} = Gp_{ij}+\tilde{E}_{j-i}p_{ij}$ if $j-i \in J$,

where

$$E_j = \int_0^1 G(x_j - t\sigma_j s_j)dt - G$$

and

$$\tilde{E}_j = \int_0^1 G(x_j - t||\sigma_j s_j|| \frac{c_{nj}}{||c_{nj}||})dt - G .$$

ii) $||D_j^{-1}P_j - G^{-1}|| = 0(\max_{i=1,\ldots,n} ||\hat{E}_{j-i}||)$,

where

$$\hat{E}_{j-i} = \begin{cases} E_{j-i} & \text{if } j-i \notin J \\ \tilde{E}_{j-i} & \text{if } j-i \in J ; \end{cases}$$

and $||\hat{E}_{j-i}|| \to 0$ as $j \to \infty$, $i = 1, \ldots, n$.

iii) There are numbers $0 < \rho_1 \leqq \rho_2$ and an integer j_0 such that, for every $x \in E^n$ and every $j \geqq j_0$,

$\rho_1||x|| \leqq \max\{|p_{ij}'x|,\ i=1,\ldots,n\} \leqq \rho_2||x||$.

Proof:

i) This part of the lemma follows immediately from Taylor's theorem and the definition of P_j and D_j.

ii) Let $W_j' = (\hat{E}_{j-1}p_{1j}, \ldots, \hat{E}_{j-n}p_{nj})$. Then $D_j = P_jG + W_j$ or $G^{-1} = D_j^{-1}P_j + D_j^{-1}W_jG^{-1}$ and it follows from Lemma 1 that

$$||D_j^{-1}P_j - G_j|| = O(\max_{i=1,\ldots,n} ||E_{j-i}||) .$$

Furthermore, by Taylor's theorem

$$F(x_j-\sigma_js_j) - F(x_j) = -\sigma_jg_j's_j+\frac{\sigma_j^2}{2}s_j'G(\xi_j)s_j$$

$$\xi_j \in E^n .$$

Because $F(x_{j+1}) \leqq F(x_j-\sigma_js_j) < F(x_j)$ and $F(x_j) - F(x_{j+1}) \to 0$ as $j \to \infty$, this implies $||\sigma_js_j|| \to 0$ and $\tilde{x}_j \to z$ as $j \to \infty$. Therefore,

$$||E_j|| \leqq ||\int_0^1 G(x_j-t\sigma_js_j)dt - G_j|| + ||G_j-G||$$

$$\leqq \sup_{0\leqq t\leqq 1}||G(x_j-t\sigma_js_j)-G_j|| + ||G_j-G|| \to 0$$

as $j \to \infty$ and

$$||\hat{E}_j|| \leqq \sup_{0\leqq t\leqq 1}||G(x_j-t||\sigma_js_j||\frac{c_{nj}}{||c_{nj}||}-G|| + ||G_j-G|| \to 0$$

as $j \to \infty$.

iii) Since $||p_{ij}|| = 1$ for all i and j, the second inequality is satisfied for every $\rho_2 \geqq 1$. From the proof of part ii) it follows that for j sufficiently large P_j^{-1} exists and

$$P_j^{-1} = (D_jG^{-1} - W_jG^{-1})^{-1} \to GD_j^{-1} \text{ as } j \to \infty .$$

Therefore, there is $\rho_1 > 0$ and j_o such that $||P_j^{-1}|| \leqq \frac{1}{\rho_1}$ for $j \geqq j_o$. This is equivalent to the first inequality.

Lemma 3

i) There is j_1 such that, for $j \geqq j_1$,

$$s_j = D_j^{-1}P_jg_j, \quad \sigma_j = \frac{g_j's_j}{2\omega_j}, \text{ and } x_{j+1} = x_{j+1} - \tilde{s}_{j+1}$$

for $j \varepsilon J$.

ii) $|1 - \sigma_j| = O(||D_j^{-1}P_j - G^{-1}|| + ||G(\eta_j) - G||)$,

where $\eta_j \varepsilon \{x | x = x_j - ts_j, \ 0 \leqq t \leqq 1\}$.

Proof:

i) By Step I of the algorithm

$$g_j'\bar{s}_j = g_j'G^{-1}g_j + g_j'(D_j^{-1}P_j - G^{-1})g_j .$$

Since $||D_j^{-1}P_j - G^{-1}|| \to 0$ and $\alpha_j \to 0$ as $j \to \infty$, we have, for j sufficiently large,

$$(1) \quad g_j'\bar{s}_j \geqq \frac{1}{2\eta}||g_j||^2 > \alpha_j||g_j||^2, \text{ i.e., } s_j = \bar{s}_j.$$

From (1) in the proof of Theorem 1 it follows that

$$(2) \quad \bar{\sigma}_j = \frac{g_j's_j}{2\omega_j} = \frac{g_js_j}{s_jG(\eta_j)s_j} ;$$

$$\eta_j \varepsilon \{x | x = x_j - ts_j, \ 0 \leqq t \leqq 1\} .$$

Since $||s_j|| = O(||g_j||)$ and, for j sufficiently large,

$$s_j'G(\eta_j)s_j = g_j's_j + s_j'G(D_j^{-1}P_j - G^{-1})g_j + s_j'(G(\eta_j) - G)s_j$$

(1) and (2) imply

(3) $|1-\bar{\sigma}_j| = O(||D_j^{-1}P_j-G^{-1}|| + ||G(\eta_j-G||)$.

By Taylor's theorem there is $\xi_j \varepsilon \{x|x=x_j-ts_j,\ 0 \leqq t \leqq 1\}$ such that, for j sufficiently large,

$$F(x_j) - F(x_j-s_j) = g_j's_j-\frac{1}{2}s_j'Gs_j-\frac{1}{2}s_j'(G(\xi_j)-G)s_j$$

$$= \frac{1}{2}g_j's_j-\frac{1}{2}s_j'G(D_j^{-1}P_j-G^{-1})g_j-\frac{1}{2}s_j'(G(\xi_j)-G)s_j.$$

Therefore, it follows from (1) that, for j sufficiently large,

(4) $\left|\frac{F(x_j)-F(x_j-s_j)}{g_j's_j} - \frac{1}{2}\right|$

$= O(||D_j^{-1}P_j-G^{-1}||+||G(\xi_j)-G||)$.

Because the right hand side of (4) goes to zero as $j \to \infty$, it follows from the definition of ν_j in Step II of the algorithm that

(5) $\sigma_j = \bar{\sigma}_j = \frac{g_js_j}{2\omega_j}$ for j sufficiently large.

In the same way it can be shown that, for $j \varepsilon J$ sufficiently large, $F(\tilde{x}_{j+1}-\tilde{s}_{j+1}) \leqq F(\tilde{x}_{j+1})$ and, therefore, $x_{j+1} = \tilde{x}_{j+1} - \tilde{s}_{j+1}$.

ii) This statement follows from (3) and (5).

Lemma 4

i) $||\tilde{g}_{j+1}|| = O(||g_j||)$, $||x_j-z|| = O(||g_j||)$,

$||g_j|| = O(||x_j-z||)$

ii) $||g_{j+1}|| = O(\|\tilde{g}_{j+1}\|)$, $||\tilde{x}_j - z|| = O(\|\tilde{g}_j\|)$,

$||\tilde{g}_j|| = O(\|\tilde{x}_j - z\|)$

Proof:

By Lagrange's formula there is
$\xi_j \in \{x \mid x = z + t(x_j - z),\ 0 \leqq t \leqq 1\}$
such that
$(x_j - z)'g_j = (x_j - z)'G(\xi_j)(x_j - z) \geqq \mu\|x_j - z\|^2$.

Thus $||x_j - z|| = O(||g_j||)$. The proof of the other statements of the lemma is similar and, therefore, omitted.

Using these results we can now prove that $\{x_j\}$ converges superlinearly to z.

Theorem 2

i) $\frac{||g_{j+1}||}{||g_j||} \to 0$ and $\frac{||x_{j+1} - z||}{||x_j - z||} \to 0$ as $j \to \infty$.

ii) If J is an infinite set, then

$$\frac{||\tilde{g}_{j+1}||}{||g_j||} \to 0,\quad \frac{||g_{j+1}||}{||\tilde{g}_{j+1}||} \to 0 \text{ as } j \to \infty,\ j \in J$$

and

$$\frac{||\tilde{x}_{j+1} - z||}{||x_j - z||} \to 0,\quad \frac{||x_{j+1} - z||}{||\tilde{x}_{j+1} - z||} \to 0 \text{ as } j \to \infty,\ j \in J.$$

Proof:

Let $j \geqq j_1$. Then it follows from Lemma 3 and Taylor's theorem that

$$\tilde{g}_{j+1} = (1-\sigma_j)g_j - \sigma_j G(D_j^{-1}P_j - G^{-1})g_j - \sigma_j\left(\int_0^1 G(x_j - t\sigma_j s_j)dt - G\right)s_j.$$

By the second part of Lemmas 2 and 3 this implies

$$\frac{||\tilde{g}_{j+1}||}{||g_j||} \to 0 \text{ as } j \to \infty.$$

Similarly, if $j \in J$, it follows that

$$g_{j+1} = -G(D_{j+1}^{-1}P_{j+1} - G^{-1})g_{j+1} - \left(\int_0^1 G\left(x_j - t||\sigma_j s_j||\frac{c_{nj}}{||c_{nj}||}\right)dt - G\right)\tilde{s}_{j+1},$$

or

$$\frac{||g_{j+1}||}{||\tilde{g}_{j+1}||} \to 0 \text{ as } j \to \infty,\ j \in J.$$

In view of Lemma 4 and the fact that $g_{j+1} = \tilde{g}_{j+1}$ if $j \notin J$, this completes the proof of the theorem.

Throughout the remainder of this section we assume now that $G(x)$ satisfies a Lipschitz condition. This assumption will be used to establish the rate of the superlinear convergence and determine the conditions under which it can be shown that J is a finite set.

Lemma 5

Suppose $j \in J$. Then

$$|p_{i,j+1}' g_{j+1}| = O(||\tilde{g}_{j+1}||\ ||g_{j-i+1}||),\ i = 1,\ldots,n$$

and

$$||g_{j+1}|| = O(||\tilde{g}_{j+1}|| \; ||g_{j-n+1}||).$$

Proof:

Let $j \geqq j_1$. Then it follows from Lemma 3 and Taylor's theorem that

$$g_{j+1} = \tilde{g}_{j+1} - GD_{j+1}^{-1}P_{j+1}\tilde{g}_{j+1} - V_j\tilde{s}_{j+1} \; ,$$

where

$$V_j = \int_0^1 G(\tilde{x}_{j+1} - t\tilde{s}_{j+1})dt - G.$$

Therefore, we have for $i = 1,\ldots,n$,

$$p_{i,j+1}'g_{j+1} = p_{i,j+1}'\tilde{g}_{j+1} - d_{i,j+1}'D_{j+1}^{-1}P_{j+1}\tilde{g}_{j+1}$$

$$- (p_{i,j+1}'G - d_{i,j+1}')D_{j+1}^{-1}P_{j+1}\tilde{g}_{j+1}$$

$$- p_{i,j+1}'V_j\tilde{s}_{j+1}$$

and

$$|p_{i,j+1}'g_{j+1}| = O((||d_{i,j+1} - Gp_{i,j+1}||$$

$$+ ||V_j||)||\tilde{g}_{j+1}||).$$

By the definition of V_j,

$$||V_j|| \leqq \sup_{0 \leqq t \leqq 1} ||G(\tilde{x}_{j+1} - t\tilde{s}_{j+1}) - G_{j+1}|| + ||G_{j+1} - G||$$

$$= O(||x_{j+1} - \tilde{x}_{j+1}|| + ||x_{j+1} - z||)$$

$$= O(||\tilde{g}_{j+1}|| + ||g_{j+1}||)$$

$$= O(||\tilde{g}_{j+1}||), \text{ (Lemma 4)}.$$

By Lemma 2,

$$(1) \quad ||d_{i,j+1} - Gp_{i,j+1}|| = \begin{cases} ||E_{j+1-i}|| & \text{if } j+1-i \notin J \\ ||\tilde{E}_{j+1-i}|| & \text{if } j+1-i \in J. \end{cases}$$

Because of Lemma 4 it follows from the definition of E_j and $\tilde{E}_j$ that, for $i = 1,\ldots,n$,

$$(2) \quad ||E_{j+1-i}|| = O(||g_{j+1-i}||) \quad \text{and}$$

$$||\tilde{E}_{j+1-i}|| = O(||g_{j+1-i}||).$$

This concludes the proof of the first statement of the lemma. Since by Lemma 2

$$\frac{||g_{j+1}||}{||\tilde{g}_{j+1}||} = O(\max\{\frac{|p'_{i,j+1}g_{j+1}|}{||\tilde{g}_{j+1}||}, \; i = 1,\ldots,n\}),$$

the second part of the lemma follows from the first statement and Lemma 4.

Lemma 6

Suppose $j \notin J$. Then

$$|1-\sigma_j| = O(||g_{j-n}||), \; |p'_{1,j+1}g_{j+1}| = O(||g_j||^2),$$

and, for $i = 1,\dots,n-1$,

$$|p'_{i+1,j+1}g_{j+1}| = |(1-\sigma_j)p'_{ij}g_j| + O(||g_j||\;||g_{j-i}||).$$

Proof:

Let $j \geqq j_1$. Combining Lemmas 3 and 4, and (1) and (2) in the proof of Lemma 5 we obtain

$$|1-\sigma_j| = O(||D_j^{-1}P_j - G^{-1}|| + ||g_j||)$$

$$= O(||g_{j-n}||).$$

Furthermore, it follows from Lemma 3 and Lagrange's formula that there is $\xi_j \in \{x \mid x = x_j - t\sigma_j s_j,\ 0 \leqq t \leqq 1\}$ such that

$$s'_j g_{j+1} = s'_j g_j - \frac{g'_j s_j}{2\omega_j} s'_j G(\xi_j) s_j.$$

Using the expression for ω_j which is given by (1) in the proof of Theorem 1 we obtain

$$g'_{j+1}s_j = g'_j s_j\left(1 - \frac{s'_j G(\xi_j)s_j}{s'_j G(\eta_j)s_j}\right),$$

$$\eta_j \in \{x \mid x = x_j - ts_j,\ 0 \leqq t \leqq 1\}.$$

Therefore,

$$|g'_{j+1}s_j| \leqq ||s_j||\;||g_j||\frac{|s'_j G(\eta_j)s_j - s'_j G(\xi_j)s_j|}{\mu||s_j||^2}$$

$$\leqq ||s_j||\;||g_j||\frac{||G(\eta_j) - G(\xi_j)||}{\mu}.$$

Since $p_{1,j+1} = s_j/||s_j||$, $||s_j|| = O(||g_j||)$, and $\{\sigma_j\}$ is bounded, this implies $|p'_{1,j+1}g_{j+1}| = O(||g_j||^2)$. By Taylor's theorem, $g_{j+1} = g_j - \sigma_j G s_j - \sigma_j U_j s_j$, where

$$U_j = \int_0^1 G(x_j - t\sigma_j s_j)dt - G.$$

Thus we have, for $i = 1,\ldots,n-1$,

$$\begin{aligned} p'_{i+1,j+1}g_{j+1} &= p'_{ij}g_{j+1} \\ &= p'_{ij}g_j - \sigma_j p_{ij} g_j \\ &\quad - \sigma_j(p'_{ij}G - d'_{ij})D_j^{-1}P_j g_j \\ &\quad - \sigma_j p'_{ij}U_j s_j \end{aligned}$$

and

$$\begin{aligned} |p'_{i+1,j+1}g_{j+1}| &= |(1-\sigma_j)p'_{ij}g_j| \\ &\quad + O((||d_{ij}-Gp_{ij}|| \\ &\qquad + ||U_j||)||g_j||). \end{aligned}$$

This completes the proof since $||d_{ij}-Gp_{ij}|| = O(||g_{j-i}||) = O(||g_{j-n}||)$.

Using the previous two lemmas we can now prove Theorem 3 which establishes the rate of the superlinear convergence of $\{x_j\}$ to z.

Theorem 3

Let $\tau_j = \max\{|p'_{ij}g_j|/||g_j||,\ i = 1,\ldots,n-1\}$.

i) If $j \notin J$, then

$$\frac{||g_{j+1}||}{||g_j||} = O(\max\{||g_{j-n+1}||, \tau_j||g_{j-n}||\}).$$

ii) If $j \in J$, then

$$\frac{||g_{j+1}||}{||g_j||} = O(||g_{j-n+1}||\max\{||g_{j-n+1}||, \tau_j||g_{j-n}||\}).$$

<u>Proof:</u>

i) By part iii) of Lemma 2,

$$\frac{||\tilde{g}_{j+1}||}{||g_j||} = O(\max\{\frac{|p_{i,j+1}'\tilde{g}_{j+1}|}{||g_j||}, i=1,\ldots,n\}).$$

Since for $j \notin J$, $g_{j+1} = \tilde{g}_{j+1}$, the statement follows from Lemmas 6 and 4.

ii) For $j \in J$, we have

$$\frac{||g_{j+1}||}{||g_j||} = \frac{||g_{j+1}||}{||\tilde{g}_{j+1}||}\,\frac{||\tilde{g}_{j+1}||}{||g_j||},$$

and, by Lemma 5 and the first part of the theorem,

$$\frac{||g_{j+1}||}{||g_j||} = O(||g_{j-n+1}||\max\{||g_{j-n+1}||, \tau_j||g_{j-n}||\}).$$

Since $|\tau_j| \leqq 1$, the above theorem states that $||g_{j+1}||/||g_j|| = O(||g_{j-n}||)$, $j \notin J$.
Assuming that $\tau_j = O(||g_{j-n+1}||/||g_{j-n}||)$ we obtain

(1) $||g_{j+1}||/||g_j|| = O(||g_{j-n+1}||)$, $j \notin J$.

We shall now show that the following result holds: If, for $j \notin J$, the actual rate of convergence is not faster than predicted by (1) and if, for $j \in J$, $||g_{j+1}||/||\tilde{g}_{j+1}||$ does not converge faster than predicted by Lemma 5, then the special updating step has to be performed at most finitely often (i.e., J is finite) and the actual rate of convergence is given by (1).

<u>Assumption 2</u>

For $j \notin J$, let $||g_{j+1}|| = \delta_j ||g_j|| \; ||g_{j-n+1}||$ and, for $j \in J$, let $||g_{j+1}|| = \beta_j ||\tilde{g}_{j+1}|| \; ||g_{j-n+1}||$. Then there is $\delta > 0$ and $\beta > 0$ such that

$$\delta_j \geqq \delta \text{ for all } j \notin J \text{ and } \beta_j \geqq \beta \text{ for all } j \in J.$$

As a first consequence of this assumption we have

<u>Lemma 7</u>

Let Assumption 2 be satisfied. Then

$$\frac{|p_{ij}' g_j|}{||g_j||} = O\left(\frac{||g_{j-i}||}{||g_{j-n}||}\right) , \quad i = 1,\ldots,n.$$

<u>Proof:</u>

If $j-1 \in J$, it follows from Lemma 5 that

$$|p_{ij}' g_j|/||g_j|| = O(||\tilde{g}_j|| \; ||g_{j-i}||/||g_j||),$$
$$i = 1,\ldots,n.$$

Since by Assumption 2, $||g_j|| \geqq \beta ||\tilde{g}_j|| \; ||g_{j-n}||$, we have

(1) $|p_{ij}'g_j|/||g_j|| = O(||g_{j-i}||/||g_{j-n}||),$

$i = 1,\ldots,n.$

If $j-1 \notin J$, it follows from Lemma 6 and Assumption 2 that

(2) $|p_{1j}'g_j|/||g_j|| = O(||g_{j-1}||/||g_{j-n}||).$

Furthermore, we conclude from Lemma 6 that for every $i \in \{2,\ldots,n\}$

(3) $$|p_{i,j-1}'g_{j-1}|/||g_{j-1}|| = O(||g_{j-1-i}||/||g_{j-1-n}||)$$

implies

(4) $$\frac{|p_{ij}'g_j|}{||g_j||} \leq \frac{|1-\sigma_{j-1}|}{||g_j||}|p_{i-1,j-1}'g_{j-1}| + O(||g_{j-i}||\frac{||g_{j-1}||}{||g_j||}$$

$$= O(||g_{j-1-n}||\frac{||g_{j-i}||}{||g_{j-1-n}||}\,\frac{||g_{j-1}||}{||g_j||}) + O(||g_{j-i}||\frac{||g_{j-1}||}{||g_j||})$$

$$= O(||g_{j-i}||/||g_{j-n}||).$$

The statement follows now from (1) - (4).

Theorem 4

Let Assumption 2 be satisfied and assume that $\gamma_2 < \mu$, where γ_2 is the constant used in Step III of the algorithm. Then

i) There is j_2 such that $j \geqq j_2$ implies $j \notin J$.

ii) $$\frac{||g_{j+1}||}{||g_j||} = O(||g_{j-n+1}||),$$

$$\frac{||x_{j+1}-z||}{||x_j-z||} = O(||x_{j-n+1}-z||), \quad j \geqq j_2$$

iii) For $j \geqq j_2$, $i = 1,\ldots,n-1$ and $k = i+1,\ldots,n$,

$$|p'_{ij}Gp_{kj}| = \left|\frac{s'_{j-i}}{||s_{j-i}||}G\frac{s_{j-k}}{||s_{j-k}||}\right|$$

$$= O(\frac{||g_{j-k}||}{||g_{j-i-n}||}).$$

Proof:

i) By Lemma 3,

(1) $$s_j = D_j^{-1}P_jg_j = \sum_{i=1}^{n} c_{ij}p'_{ij}g_j \quad \text{for } j \geqq j_1.$$

Let $||s_j|| = \rho_j||g_j||$. Since $D_j^{-1}P_j \to G^{-1}$ as $j \to \infty$, there is $\rho > 0$ such that $\rho_j \geqq \rho > 0$ for j sufficiently large. Observing that $||g_{j-i}||/||g_{j-n}|| \to 0$ as $j \to \infty$, $i = 1,\ldots,n-1$, we conclude from (1) and Lemma 7 that

(2) $$\frac{s_j}{||s_j||} = \sum_{i=1}^{n} c_{ij}\frac{p'_{ij}g_j}{\rho_j||g_j||} \to c_{nj}\frac{p'_{nj}g_j}{\rho_j||g_j||}$$

as $j \to \infty$.

By Lagrange's formula there is

$\xi_j \in \{x \mid x = x_j - t\sigma_j s_j,\ 0 \leq t \leq 1\}$ such that

$$c'_{nj}d_j = \frac{c'_{nj}(g_j - g_{j+1})}{||\sigma_j s_j||} = c'_{nj}G(\xi_j)\frac{s_j}{||s_j||}$$

$$= c'_{nj}G(\xi_j)\frac{c_{nj}}{||c_{nj}||} + c'_{nj}G(\xi_j)\left(\frac{s_j}{||s_j||} - \frac{c_{nj}}{||c_{nj}||}\right)$$

$$\geq \mu||c_{nj}|| - \eta||c_{nj}||\,\left|\left|\frac{s_j}{||s_j||} - \frac{c_{nj}}{||c_{nj}||}\right|\right| .$$

Therefore, it follows from (2) that $c'_{nj}d_j \geq \gamma_2||c_{nj}||$, i.e., $j \notin J$, for j sufficiently large.

ii) By Lemma 7 , we have for j sufficiently large,

$$\tau_j = \max\left\{ \frac{|p'_{ij}g_j|}{||g_j||},\ i = 1,\ldots,n-1\right\}$$

$$= O\left(\frac{||g_{j-n-1}||}{||g_{j-n}||}\right).$$

Therefore, it follows from Theorem 3 that $||g_{j+1}|| = O(||g_j||\,||g_{j-n+1}||)$ which, by Lemma 4, implies $||x_{j+1}-z|| = O(||x_j-z||\,||x_{j-n+1}-z||)$.

iii) It follows from Lemma 2 that, for $j \geq j_2+n$ and $i = 1,\ldots,n$,

$$(3)\qquad \begin{aligned} d_{ij} &= Gp_{ij} - E_{j-i}p_{ij} \quad \text{and} \\ p_{ij} &= p_{j-i} = s_{j-i}/||s_{j-i}|| \end{aligned}$$

where, by (2) in the proof of Lemma 5, $||E_{j-i}|| = O(||g_{j-i}||)$. Furthermore, for j sufficiently large,

$$p'_{ij}G\frac{s_j}{||s_j||} = p'_{ij}GD_j^{-1}P_j\frac{g_j}{||s_j||}$$

$$= \frac{p'_{ij}g_j}{||s_j||} - p'_{ij}E_{j-i}\frac{s_j}{||s_j||} \quad .$$

Therefore, Lemma 7 implies that, with $p_j = s_j/||s_j||$,

(4) $|p'_jGp_{ij}| = O(||g_{j-i}||/||g_{j-n}||)$, $i=1,\ldots,n$.

For $j \geqq j_2+n$ it follows from (3) and the definition of P_j that $p_{ij} = p_{j-i}$ and $p_{kj} = p_{k-i,j-i}$, $i = 1,\ldots,n-1$, $k = i+1,\ldots,n$. Applying (4) with j replaced by j-i we obtain, therefore, for $i = 1,\ldots,n-1$, $k = i+1,\ldots,n$,

$$|p'_{ij}Gp_{kj}| = |p_{j-i}Gp_{k-i,j-i}|$$

$$= O(\frac{||g_{j-i-(k-i)}||}{||g_{j-i-n}||})$$

$$= O(\frac{||g_{j-k}||}{||g_{j-i-n}||}).$$

6. A modified algorithm and its convergence properties

Since it is in general difficult to verify assumptions on second order derivatives of F(x) we shall consider a modification of Algorithm 1 for which convergence can be established if F(x) is only assumed to be continuously differentiable.

In addition, we shall show that the modified algorithm has the same superlinear convergence properties as the original algorithm if a rather weak assumption on second order derivatives of $F(x)$ is satisfied.

Throughout this section we shall make the following

<u>Assumption 3:</u>

The modified algorithm starts with an x_0 such that

$$S_0 = \{x \mid F(x) \leqq F(x_0)\}$$

is bounded and $F(x)$ is continuously differentiable on some convex open set S containing S_0.

It is clear that Assumption 1 implies that Assumption 3 holds for any $x_0 \varepsilon E^n$. It suffices, therefore, to modify those parts of Algorithm 1 which depend on properties of the Hessian matrix of $F(x)$.

This is the case in Step II in the definition of $\bar{\sigma}_j$. Under Assumption 3 it cannot be guaranteed that $\omega_j > 0$ or that $\{\bar{\sigma}_j\}$ is bounded and bounded away from zero. Therefore, a test is introduced which ensures $\bar{\sigma}_j = g_j' s_j / 2\omega_j$ if and only if $\omega_j > 0$ and $g_j' s_j / 2\omega_j$ is in an appropriately chosen interval.

In proving the important Lemma 1 we used that the sequences $\{d_j\}$ and $\{d_j\}$ are bounded and that $|d_j c_{nj}| \geqq \mu \|c_{nj}\|$ for all j. Again Assumption 3 is not strong enough to ensure these properties and we have to modify Steps III and IV by introducing additional tests.

We described now a general cycle of the modified algorithm. At the beginning of the j-th cycle the following information is available: $x_j \in E^n$, g_j, D_j^{-1}, and P_j.

<u>Algorithm 2</u>

<u>Step I</u>: <u>Determination of s_j</u>

Same as in Algorithm 1

<u>Step II</u>: <u>Determination of σ_j</u>

Compute $\omega_j = F(x_j - s_j) - F(x_j) + g_j' s_j$ and set

$$\bar{\sigma}_j = \begin{cases} \delta_1 & \text{if } \omega_j > 0 \text{ and } \dfrac{g_j' s_j}{2\omega_j} \leq \delta_1 \\ \dfrac{g_j' s_j}{2\omega_j} & \text{if } \omega_j > 0 \text{ and } \delta_1 < \dfrac{g_j' s_j}{2\omega_j} < \delta_2 \\ \delta_2 & \text{if } \omega_j \leq 0 \text{ or } \omega_j > 0 \text{ and } \dfrac{g_j' s_j}{2\omega_j} \geq \delta_2, \end{cases}$$

where $0 < \delta_1 < 1 < \delta_2$. Continue as in Algorithm 1.

<u>Step III</u>: <u>Updating of D_j^{-1} and P_j</u>

Let

$$d_j = \frac{g_j - g_{j+1}}{||\sigma_j s_j||} \quad \text{and} \quad p_j = \frac{s_j}{||s_j||}$$

<u>Case 1:</u> $|d_j' c_{nj}| \geq \gamma_2 ||c_{nj}||$ and $||d_j|| \leq \gamma_3$, $\gamma_2 > 0$, $\gamma_3 > 0$.

Continue as in Algorithm 1.

Case 2: $|d_j'c_{nj}| \geqq \gamma_2||c_{nj}||$ and $||d_j|| > \gamma_3$.

Set $D_{j+1}^{-1} = \Phi(D_j^{-1}, \frac{c_{nj}}{||c_{nj}||})$,

$P_{j+1} = \Psi(P_j, \frac{c_{nj}}{||c_{nj}||})$, $x_{j+1} = x_{j+1}$ and $g_{j+1} = g_{j+1}$.

Replace j by j+1 and go to Step I.

Case 3: $|d_j'c_{nj}| < \gamma_2||c_{nj}||$ and $||d_j|| \leqq \gamma_3$.

Go to Step IV.

Step IV: Special updating step

Compute

$$d_j = \frac{g_j - \nabla F(x_j - ||\sigma_j s_j|| \frac{c_{nj}}{||c_{nj}||})}{||\sigma_j s_j||}$$

Set $P_{j+1} = \Psi(P_j, \frac{c_{nj}}{||c_{nj}||})$ and

$$D_{j+1}^{-1} = \begin{cases} \Phi(D_j^{-1}, d_j) \text{ if } |d_j'c_{nj}| \geqq \gamma_4||c_{nj}|| \text{ and } ||d_j|| \leqq \gamma_3 \\ \Phi(D_j^{-1}, \frac{c_{nj}}{||c_{nj}||}) \text{ otherwise,} \end{cases}$$

where $\gamma_4 > 0$. Continue as in Algorithm 1.

First we note that Lemma 1 applies also to Algorithm 2. Indeed, a review of the proof of Lemma 1 shows that the statements are true if

i) $D_{j+1}^{-1} = \Phi(D_j^{-1}, u_j)$, $P_{j+1} = \Psi(P_j, v_j)$ implies $||u_j|| \leqq \rho_1$ and $|u_j'v_j| \geqq \rho_2||v_j||$ for some $\rho_1, \rho_2 > 0$.

ii) $\{g_j\}$ is bounded.

The first condition is clearly satisfied and $\{g_j\}$ is bounded because $\{x_j\} \subset S$.

Theorem 4

Let Assumption 3 be satisfied. Then Algorithm 2 either terminates after a finite number of steps with a stationary point or generates an infinite sequence $\{x_j\}$ with the properties

i) $F(x_{j+1}) < F(x_j)$ for all j
ii) $||g_j|| \to 0$, $||x_{j+1}-x_j|| \to 0$ as $j \to \infty$
iii) Every cluster point of $\{x_j\}$ is a stationary point
iv) If $\{x_j\}$ has an isolated cluster point z, then $x_j \to z$ as $j \to \infty$.

Proof:

It follows immediately from the definition of Algorithm 2 that $F(x_{j+1}) < F(x_j)$ for every j and that the algorithm terminates with x_j if and only if $\nabla F(x_j) = 0$. If Algorithm 2 generates an infinite sequence $\{x_j\}$, then it follows as in the proof of Theorem 1 that $||g_j|| \to 0$ as $j \to \infty$. Since $\{D_j^{-1}P_j\}$ is bounded and $\sigma_j \leqq \delta_2$ for all j this implies $||x_{j+1}-x_j|| \to 0$ as $j \to \infty$. If z is a cluster point of $\{x_j\}$, it follows from the continuity of $\nabla F(x)$ that $\nabla F(z) = 0$. Finally, the last statement of the theorem is a well-known consequence of the fact that $||x_{j+1}-x_j|| \to 0$ as $j \to \infty$.

In order to prove that $\{x_j\}$ converges superlinearly to a (local) minimizer of $F(x)$ we need some assumptions on second order derivatives of $F(x)$. These are stated in

Assumption 4

i) The sequence $\{x_j\}$ generated by Algorithm 2 has a cluster point z with the following properties: $F(x)$ is twice continuously differentiable in some convex neighborhood of z and there are numbers $0 < \bar{\mu} < \bar{\eta}$ such that $\bar{\mu}||x||^2 \leqq x'G(z)x \leqq \bar{\eta}||x||^2$ for all $x \in E^n$.

ii) $\gamma_2 < \bar{\mu}$, $\gamma_4 < \bar{\mu}$, and $\gamma_3 > \bar{\eta}$, where γ_i are the constants used in Steps III and IV of Algorithm 2.

Lemma 8

Let Assumptions 3 and 4 be satisfied. Then

i) $x_j \to z$ as $j \to \infty$.
ii) There is j_3 such that, for $j \geqq j_3$,

$$|d_j'c_{nj}| \geqq \gamma_2||c_{nj}||, \quad ||d_j|| \leqq \gamma_3,$$

and

$$|\tilde{d}_jc_{nj}| \geqq \gamma_4||c_{nj}||, \quad ||\tilde{d}_j|| \leqq \gamma_3.$$

Proof:

The first statement follows from Theorem 4 since Assumption 4 implies that z is an isolated cluster point of $\{x_j\}$. The proof of part ii) is an easy application of Lagrange's formula and, therefore, omitted.

Let Assumption 4 be satisfied. Then there is a neighborhood U of z in which F(x) has the same properties as under Assumption 1. Furthermore, Lemma 8 implies that, for j sufficiently large, the Algorithms 1 and 2 are equivalent and $x_j \in U$. Therefore, we have

Theorem 5

Let Assumptions 3 and 4 be satisfied. Then the Theorems 3 and 4 apply also to the sequences $\{x_j\}$ and $\{g_j\}$ generated by Algorithm 2.

References

[1] Armijo,L.,"Minimization of functions having Lipschitz continuous first partial derivatives", Pacific Journal of Mathematics, 16, 1-3, 1966.

[2] Davidon,W.C.,"Variable metric method for minimization", A.E.C.Development Report, ANL-5990, 1959.

[3] Dieudonné,J., "Foundations of modern analysis", Academic Press, New York, 1960.

[4] Fletcher, R., and Powell, M.J.D., "A rapidly convergent descent method for minimization", The Computer Journal, 6 , 163-168, 1963.

[5] Goldstein, A.A., "Constructive real analysis", Harper and Row, New York, 1967.

[6] Goldstein, A.A., and Price, J.F., "An effective algorithm for minimization", Numerische Mathematik, 10, 184-189, 1967.

[7] McCormick, G.P., and Ritter, K., "Methods of conjugate directions versus quasi-Newton methods", Mathematical Programming, 3, 101-116, 1972.

[8] Powell, M.J.D., "On the convergence of the variable metric algorithm", J.Inst.Maths.Applcs., 7, 21-36, 1971.

[9] Ritter, K., "A superlinearly convergent method for unconstrained minimization", Nonlinear Programming, J.B.Rosen, O.L.Mangasarian, and K. Ritter, editors, Academic Press, New York, 1970.

[10] Ritter, K., "Superlinearly convergent methods for unconstrained minimization problems", Proc. ACM, 1137-1145, 1972.

[11] Vainberg, M.M., "Variational methods for the study of nonlinear operators", Holden-Day, San Francisco, 1964.

SUPERLINEARLY CONVERGENT ALGORITHMS FOR LINEARLY CONSTRAINED OPTIMIZATION PROBLEMS[1)]

by

U. M. García-Palomares[2)]

ABSTRACT

In this paper new algorithms for solving linearly constrained optimization problems are proposed. It is shown that certain updating schemes which have been successfully used in unconstrained optimization can also be used to implement these new algorithms. It is proved, that, under suitable conditions the sequence of points generated by the algorithms converges Q-superlinearly from any initial feasible point to a stationary point, that is, a point which satisfies the Kuhn-Tucker conditions. The subproblems solved by the algorithms are minimization problems with a quadratic objective function and linear constraints. The constraints are the ε-active constraints of the original problem at the iteration point.

1. Introduction

The algorithms introduced in this paper are feasible directions algorithms, in the sense that a

---

[1] Supported by NSF Grant GJ35292
[2] Instituto Venezolano de Investigaciones Científicas (I.V.I.C.) Apartado 1827, Caracas, Venezuela

sequence of estimates to the solution is generated by determining at each iteration a direction which is bounded, feasible and usable, as defined in [Topkis and Veinott 67]. An important feature of the algorithms is the use of updating schemes and the use of ε-active constraints. It is proved in section 3 that under suitable conditions the iterates converge Q-superlinearly [Ortega and Rheinboldt 70] to a stationary point.

Superscripts will denote components of a vector or columns of a matrix, subscripts will denote iteration number or members of a sequence, and $\|\cdot\|$ will denote an arbitrary, but fixed norm. All vectors w will be column vectors unless transposed to a row by a superscript T.

2. The Algorithm

The algorithm will solve the following linearly constrained problem

$$\text{(LCP)} \quad \underset{x\in\psi}{\text{minimize}}\ f(x) \qquad \psi = \{x \mid x^T A \leq b^T,\ x \in R^n\}$$

where $f: R^n \to R$, A is an $n\times m$ matrix, and $b \in R^m$.

We assume that we are given some $\varepsilon > 0$, an initial feasible point x_0 and a positive definite matrix G_0. Given x_i, and a positive definite matrix G_i, the algorithm proceeds as follows to obtain x_{i+1} and G_{i+1}:

Step 1: (Direction Finding): Solve the following convex quadratic problem (Q_i) and denote its solution by p_i

$$(Q_i) \quad \underset{p \in \psi_i}{\text{minimize}} \quad \nabla^T f(x_i)p + \tfrac{1}{2}p^T G_i p$$

$$\psi_i = \{p \mid p^T A_i \leqq b_i^T - x_i^T A_i\}$$

where A_i is a submatrix of A formed by the columns A^j, $j \in \Gamma(x_i) = \{j \mid j \in \{1,2,\ldots,m\},\ -\varepsilon \leqq x_i^T A^j - b^j \leqq 0\}$, and b_i is a vector formed by the components b^j, $j \in \Gamma(x_i)$.

Step 2 (Feasibility): Pick

$$\mu_i = \text{minimum} \left\{ 1, \frac{b^j - x_i^T A^j}{p_i^T A^j} \right\} \text{ for } j \notin \Gamma(x_i) \text{ and } p_i^T A^j > 0$$

Step 3 (Stepsize): If some convergence criterion is satisfied, such as $\|p_i\| \leqq \delta$ for some small given $\delta > 0$, terminate. Otherwise choose any stepsize procedure, $x_{i+1} = x_i + \lambda_i \mu_i p_i$, with the property that $\lim_{\substack{i \to \infty \\ i \in \Lambda}} \mu_i \nabla^T f(x_i) p_i = 0$ for $\Lambda \subset \{1,2,\ldots,\}$ such that $\{x_i\}_{i \in \Lambda} \to \bar{x}$.

Step 4 (Updating): Update the matrix G_i such that each matrix in the sequence of matrices $\{G_i\}$ is a bounded and uniformly positive definite $n \times n$ matrix, that is, $\nu_2 \|p\|^2 \geqq p^T G_i p \geqq \nu_1 \|p\|^2$ for some $\nu_2 \geqq \nu_1 > 0$ and all $p \in R^n$.

Remark 2.1: The subproblems (Q_i) can be solved by finite and efficient methods [Cottle and Dantzig 68, Stoer 71].

Remark 2.2: The only requirements on G_i is boundedness and positive definiteness, hence the work by [Pironneau and Polak 71] is a particular case, namely $G_i = I$ the identity matrix. They showed a linear rate of convergence.

Remark 2.3: The dual of (Q_i) is the subproblem of the dual, feasible direction algorithm [Mangasarian 72], therefore the sequence $\{x_i\}$ generated by both algorithms is the same. However, the superlinear rate of convergence is given here for the first time.

Remark 2.4: Several stepsize procedures can be used that satisfy the property given in step 3 [Daniel 70]. The Armijo procedure [Armijo 66] has been used for its simplicity. For completeness we give the Armijo procedure below:

For some $\alpha \in (0,1)$ and $\beta \in (0,\frac{1}{2})$ pick

λ_i = Maximum $\{1,\alpha,\alpha^2,\ldots,\alpha^n,\ldots\}$ such that

$$f(x_i+\lambda_i\mu_i p_i) - f(x_i) \leqq \beta\lambda_i\mu_i\nabla^T f(x_i)p_i$$

and define

$$x_{i+1} := x_i + \lambda_i\mu_i p_i \ .$$

3. Analysis and Convergence of the Algorithm

Before we give the convergence results we need to establish several lemmas:

Lemma 3.1 ($\{u_i\}$ bounded) If the interior of the set ψ is nonempty, that is,

$$\text{Tnt } (\psi) = \{x \mid x^T A < b^T, x \in R^n\} \neq \phi$$

and $\nabla f(x_i)$, G_i, and p_i are all bounded, then the sequence of multipliers $\{u_i\}$ associated with problems $\{Q_i\}$ is bounded.

Proof: Let u_i be the multiplier associated with problem (Q_i). For every $j \in \Gamma(x_i)$ such that $p_i^T A^j = b_i^j - x_i^T A^j$, and $u_i^j > 0$, then $(x - x_i - p_i)^T A^j < 0$ for $x \in \text{Int } (\psi)$. Let B_i be the matrix formed by A^j such that $j \in \Gamma(x_i)$ and $u_i^j > 0$. Then $(x - x_i - p_i)^T B_i < 0$ and by Gordan's theorem there exists no $z \neq 0$, $z \geqq 0$ such that $B_i z = 0$. Hence

$$\underset{z \geqq 0,\ \|z\| = 1}{\text{minimum}} \quad z^T B_i^T B_i z > 0$$

Since the number of constraints is finite we have that $u^T B_i^T B_i u \geqq \alpha \| u \|^2$ for some $\alpha > 0$. Hence

$$\| u_i \|^2 \leqq \frac{1}{\alpha} \| B_i u_i \|^2 = \frac{1}{\alpha} \| \nabla f(x_i) + G_i p_i \|^2$$

which is bounded by assumption.

Lemma 3.2 ($\{p_i\}$ bounded and $1 \geqq \mu_i \geqq \bar{\mu} > 0$) Let the following assumptions hold:

a. The set $\{\nabla f(x_i)\}$ is bounded

b. $\{G_i\}$ is a sequence of bounded and uniformly positive definite matrices.

Then the sequence $\{p_i\}$ is bounded,

$$\nabla^T f(x_i)p_i \leq -\frac{\nu_1}{2}\|p_i\|^2 \quad \text{and} \quad 1 \geq \mu_i \geq \bar{\mu} > 0$$

Proof: $\nabla^T f(x_i)p_i + \frac{1}{2}p_i^T G_i p_i \leq 0$ (because $p = 0$ is feasible)

Hence $\nabla^T f(x_i)p_i \leq -\frac{1}{2}p_i^T G_i p_i \leq -\frac{\nu_1}{2}\|p_i\|^2$

But since $-\|\nabla f(x_i)\| \, \|p_i\| \leq \nabla^T f(x_i)p_i$, we get that $-\|\nabla f(x_i)\| \, \|p_i\| \leq -\nu_1 \|p_i\|^2$ and hence

$\|p_i\| \leq \dfrac{\|\nabla f(x_i)\|}{\nu_1}$. The boundedness of $\{p_i\}$ then follows from the boundedness of $\{\nabla f(x^i)\}$.

Since $\mu_i = \text{minimum}\,\{1, \dfrac{b^j - x_i^T A^j}{p_i^T A^j}\}$, $j \notin \Gamma(x_i)$, then

$$\mu_i \geq \text{minimum}\,\{1, \frac{\varepsilon}{p_i^T A^j}\}$$

$$\geq \text{minimum}\,\{1, \frac{\nu_1 \varepsilon}{\|\nabla f(x_i)\| \, \|A^j\|}\}$$

$\geq \text{minimum}\,\{1, \frac{\nu_1 \varepsilon}{\gamma\delta}\} = \bar{\mu}$, where $\|\nabla f(x_i)\| \leq \gamma$ and $\max\limits_{1 \leq j \leq m} \|A^j\| \leq \delta$.

Lemma 3.3 (Kuhn-Tucker conditions). Let $\Lambda \subset \{1,2,\ldots,n,\ldots\}$ and let the following assumptions hold:

a. $\{x_i\}_{i\in\Lambda} \to \bar{x}$, $\{p_i\}_{i\in\Lambda} \to 0$

b. $\{x \mid x^T A < b^T, x \in R^n\} \neq \phi$

c. $f \in C^1$, and $\{G^i\}$ is bounded

Then $\bar{x}$ and some $\bar{u}$ is a Kuhn-Tucker point for (LCP).

<u>Proof</u>: The Kuhn-Tucker conditions for (Q) are

$$p_i^T A_i \leqq b_i^T - x_i^T A_i$$

$$(p_i^T A_i - b_i^T + x_i^T A_i) u_i = 0$$

$$\nabla f(x_i) + G_i p_i + A_i u_i = 0$$

$$u_i \geqq 0$$

Since the number of constraints is finite there is some $\bar{i}$ such that $A_{\bar{i}}$ and $b_{\bar{i}}$ appear an infinite number of times in the sequence $\{(A_i, b_i)\}$. Hence we can extract a further subsequence Λ_1 for which $\{(A_{\bar{i}}, b_{\bar{i}})\} = \{(A_i, b_i)\}$ $i \in \Lambda$, and by lemma 3.1 $\{u_i\}_{\Lambda_1} \to \bar{u}$. Hence, in the limit we have that:

$$0 \leqq b_{\bar{i}}^T - \bar{x}^T A_{\bar{i}}$$

$$(-b_{\bar{i}}^T + \bar{x}^T A_{\bar{i}}) \bar{u} = 0$$

$$\nabla f(\bar{x}) + A_{\bar{i}} \bar{u} = 0$$

$$\bar{u} \geqq 0$$

and by defining $\bar{u}^j = 0$ for $j \notin \Gamma(\bar{x})$ we conclude that $(\bar{x}, \bar{u})$ satisfies the Kuhn-Tucker conditions for (LCP).

We are now ready to give precise sufficient conditions for the convergence of the algorithm.

<u>Theorem 3.1</u> (Convergence theorem) Assume that

a. $f \in C^1$

b. $\{x | f(x) \leqq f(x_o), x \in \psi\}$ is bounded

c. $\{G_i\}$ is a sequence of bounded and uniformly positive definite matrices

d. $\{x | x^T A < b^T, x \in R^n\} \neq \phi$

Then for any accumulation point $\bar{x}$ of $\{x_i\}$ there exists some $\bar{u}$, such that $(\bar{x},\bar{u})$ satisfies the Kuhn-Tucker conditions for (LCP)and $\{p_i\}_{i \in \Lambda} \to 0$ for $\{x_i\}_{i \in \Lambda} \to \bar{x}$.

<u>Proof</u>: By lemma 3.2, $1 \geqq \mu_i \geqq \bar{\mu} > 0$. Take $\Lambda \subset \{1,2,...\}$ such that $\{x_i\}_{i \in \Lambda} \to \bar{x}$. Then for $i \in \Lambda$ we have that

$$
\begin{aligned}
0 &= \lim_{i \to \infty} \mu_i \nabla^T f(x_i) p_i \quad \text{(by step 3 of the algorithm)} \\
&\leqq \bar{\mu} \lim_{i \to \infty} \nabla^T f(x_i) p_i \quad (\text{since } \mu_i \geqq \bar{\mu} > 0) \\
&\leqq \bar{\mu} \lim_{i \to \infty} \left(- \frac{\nu_1}{2} \|p_i\|^2\right) \quad \text{(by lemma 3.2)} \\
&\leqq 0
\end{aligned}
$$

and hence we conclude that $\{p_i\}_{i \in \Lambda} \to 0$, and by lemma 3.3 there exists a $\bar{u}$ such that $(\bar{x},\bar{u})$ satisfies the Kuhn-Tucker conditions for (LCP).

To establish a superlinear rate of convergence we have to impose a somewhat more stringent conditions. Step 3 and step 4 of the algorithm are modified to the following:

Step 3' The stepsize procedure satisfies the following property: $x_{i+1} = x_i + \lambda_i \mu_i p_i$ and $\{\lambda_i\} \to 1$

Step 4' The updating scheme satisfies the following condition:

$$\lim_{i\to\infty} \frac{\|(G_i - \nabla^2 f(x_i))p_i\|}{\|p_i\|} = 0$$

Remark 3.1 The additional conditions required above are necessary and sufficient for Q-superlinear rate of convergence in unconstrained optimization [Dennis and More 74]. Here we assert that these conditions are also sufficient for Q-superlinear rate of convergence in the linearly constrained case.

We first establish some lemmas.

Lemma 3.4 ($\{x_i\} \to \bar{x}$) If $\{x_i\}$ is a bounded sequence in R^n which does not have a continuum of accumulation points, and if for each convergent subsequence $\{x_i\}_{i\in\Lambda}$ it follows that $\{x_{i+1} - x_i\}_{i\in\Lambda} \to 0$, then $\{x_i\}$ converges.

Proof: We first show that $\{x_{i+1} - x_i\} \to 0$ for the entire sequence. For suppose not. Then for some convergent subsequence $\{x_i\}_{i\in\Lambda_1}$,

$\|x_{i+1} - x_i\| \geq \delta > 0$ for some $\delta > 0$ which contradicts the hypothesis of the lemma. Hence $\{x_{i+1} - x_i\} \to 0$ and the lemma follows from Ostrowski's theorem [Ostrowski 66, Daniel 71] which states that a bounded sequence $\{x_i\}$ in R^n for which $\{x_{i+1}-x_i\} \to 0$ and which does not have a continuum of accumulation points converges.

<u>Lemma 3.5</u> If $\{p_i\} \to 0$ then $\mu_i = 1$ for all i large enough and $x_i \in \psi$.

<u>Proof</u>: Since $\mu_i = \underset{j \notin \Gamma(x_i)}{\text{minimum}} \{1, \frac{b^j - x_i^T A^j}{p_i^T A_j}\}$, then

$\mu_i \geqq \text{minimum} \{1, \frac{\varepsilon}{p_i^T A_j}\}$. If $\{p_i\} \to 0$, then for large enough i, $\frac{\varepsilon}{p_i^T A_j} \geqq 1$ and therefore $\mu_i = 1$ and $x_i \in \psi$.

<u>Lemma 3.6</u> ($\lambda_i = 1$ for large enough i) Let the following assumptions hold:

a. $\{x_i\} \to \bar{x}$ and $\{p_i\} \to 0$

b. $f \in C^2$ in $N(\bar{x}, \delta)$ for some $\delta > 0$, where $N(\bar{x}, \delta) := \{x \mid \|x-\bar{x}\| < \delta\}$

c. $\lim_{i \to \infty} \frac{p_i^T (\nabla^2 f(x_i) - G_i) p_i}{\|p_i\|^2} = 0$

Then for the Amijo procedure $\lambda_i = 1$ for i large enough.

Proof: By lemma 3.5 $\mu_i = 1$ for all i large enough. Take i large enough so that $y_i := x_i + p_i$ belongs to $N(\bar{x},\delta)$. We have that

$$f(y_i) = f(x_i) + \nabla^T f(x_i)p_i + \frac{1}{2}p_i^T \nabla^2 f(z_i)p_i \quad \text{for some} \quad z_i \in (x_i,y_i)$$

hence

$$\frac{f(y_i) - f(x_i)}{\nabla^T f(x_i)p_i} = 1 + \frac{1}{2}\,\frac{p_i^T \nabla^2 f(z_i)p_i}{\nabla^T f(x_i)p_i}$$

If $p_i^T \nabla^2 f(z_i)p_i \leqq 0$ then $\lambda_i = 1$, because $\nabla^T f(x_i)p_i \leqq \beta\nabla f(x_i)p_i$. So let us assume that $p_i^T \nabla^2 f(z_i)p_i > 0$. Then

$$\frac{f(y_i) - f(x_i)}{\nabla^T f(x_i)p_i} \geqq \frac{1}{2} - \frac{1}{2}\,\frac{p_i^T(\nabla^2 f(z_i) - G_i)p_i}{p_i^T G_i p_i}$$

(because $\nabla^T f(x_i)p_i \leqq -p_i^T G_i p_i$ by the minimum principle)

$$\geqq \frac{1}{2} - \frac{1}{2\nu_1}\,\frac{p_i^T(\nabla^2 f(z_i) - G_i)p_i}{\|p_i\|^2}$$

$$= \frac{1}{2} - \frac{1}{2\nu_1}\,\frac{p_i^T((\nabla^2 f(z_i) - \nabla^2 f(x_i)) + \nabla^2 f(x_i) - G_i)p_i}{\|p_i\|^2}$$

Since $z_i \in (y_i,x_i)$ we conclude by assumptions b and c that

$$\lim_{i\to\infty} \frac{f(y_i) - f(x_i)}{\nabla^T f(x_i) p_i} \geq \frac{1}{2} > \beta$$

and hence the Armijo stepsize condition holds for $\lambda_i = 1$ if i is large enough.

Lemma 3.7 If $\lim_{i\to\infty} \frac{\|x_{i+1}-x_i\|}{\|x_i-x_{i-1}\|} = 0$, then $\{x_i\} \to \bar{x}$ and

$$\lim_{i\to\infty} \frac{\|x_{i+1}-\bar{x}\|}{\|x_i-\bar{x}\|} = 0 .$$

Proof: By assumption, $\|x_{i+1}-x_i\| \leq \varepsilon_i \|x_i-x_{i-1}\|$ and $\lim_{i\to\infty} \varepsilon_i = 0$. Choose $\bar{i}$ such that $\varepsilon_i \leq \gamma < 1$ for some positive γ and all $i > \bar{i}$. If we define $c = (\varepsilon_1 \varepsilon_2, \ldots, \varepsilon_{\bar{i}})\gamma^{-\bar{i}} \|x_1-x_0\|$, then for $i > \bar{i}$

$$\|x_{i+1}-x_i\| \leq \varepsilon_i \varepsilon_{i-1}, \ldots, \varepsilon_1 \|x_1-x_0\| \leq c\gamma^i$$

and for $k > i > \bar{i}$

$$\|x_k-x_i\| \leq \|x_k-x_{k-1}\| + \cdots + \|x_{i+1}-x_i\|$$
$$\leq c(\gamma^{k-1}+\cdots+\gamma^i) \leq \frac{c\gamma^i}{1-\gamma}$$

from which it follows that $\|x_k-x_i\| \to 0$ as $k,i \to \infty$ and hence $\{x_i\}$ is a Cauchy sequence which converges to some $\bar{x}$. Now for every $\varepsilon \in (0,1)$ there exist an $i(\varepsilon)$ such that $\frac{\|x_{i+1}-x_i\|}{\|x_i-x_{i-1}\|} \leq \varepsilon$

for all $i \geq i(\varepsilon)$. Hence for $i \geq i(\varepsilon)$

$$\begin{aligned}\|x_{i+1}-\bar{x}\| &\leq \|x_{i+1}-x_{i+2}\| + \|x_{i+2}-x_{i+3}\| + \cdots \\ &\leq \|x_i-x_{i+1}\| (\varepsilon+\varepsilon^2+\cdots) = \\ &= \frac{\varepsilon}{1-\varepsilon}\|x_i-x_{i+1}\|\end{aligned}$$

and

$$\begin{aligned}\|x_i-\bar{x}\| &\geq \|x_i-x_{i+1}\| - \|x_{i+1}-x_{i+2}\| - \cdots\cdots\cdot \\ &\geq \|x_i-x_{i+1}\| (1-\varepsilon-\varepsilon^2-\varepsilon^3\cdots) \\ &= \frac{(1-2\varepsilon)}{1-\varepsilon}\|x_i-x_{i+1}\|\end{aligned}$$

Hence

$$\frac{\|x_{i+1}-\bar{x}\|}{\|x_i-\bar{x}\|} \leq \frac{\varepsilon}{1-2\varepsilon}$$

Since $\varepsilon \in (0,1)$ was arbitrary, we have that

$$\lim_{i\to\infty} \frac{\|x_{i+1}-\bar{x}\|}{\|x_i-\bar{x}\|} = 0 .$$

We are now ready to establish sufficient conditions for a Q-superlinear rate of convergence.

Theorem 3.2 (Superlinear Rate of Convergence). In addition to the assumptions of theorem 3.1 let $f \in C^2$ and Lipschitz continuous in $N(\bar{x},\delta)$, that is, $\|\nabla^2 f(y) - \nabla^2 f(x)\| \leq \eta \|y-x\|$ for all $x,y \in N(\bar{x},\delta)$ and some $\eta > 0$. Let

$$\lim_{i\to\infty} \frac{\|(G_i-\nabla^2 f(x_i))p_i\|}{\|p_i\|} = 0 ,$$

and let (LCP) have no continuum of Kuhn-Tucker points in R^n. Then the sequence $\{x_i\}$ converges Q-superlinearly to a Kuhn-Tucker point.

Proof: By theorem 3.1, each accumulation point of $\{x_i\}$ satisfies the Kuhn-Tucker conditions of (LCP) and $\{p_i\}_{i\in\Lambda} \to 0$ for each convergent subsequence $\{x_i\}_{i\in\Lambda}$. Hence by lemma 3.4 it follows that $\{x_i\}$ converges to a Kuhn-Tucker point $\bar{x}$ of (LCP), and again by theorem 3.1 we have that $\{p_i\} \to 0$. By lemmas 3.5 and 3.6 we have that $\mu_i = 1$ and $\lambda_i = 1$ for i large enough. Hence assume that i is large enough, so that $x_{i+1} = x_i + p_i$ and $x_i \in \psi$. For convenience we let $\nabla f_i := \nabla f(x_i)$ and so on.

$$-\frac{\nu_1}{2}\|p_{i+1}\|^2 \geq \nabla^T f_{i+1} p_{i+1} \quad \text{(by lemma 3.2)}$$

$$= (\nabla f_{i+1} - \nabla f_i - \nabla^2 f_i p_i)^T p_{i+1}$$

$$- [(G_i - \nabla^2 f_i)p_i]^T p_{i+1} + (\nabla f_i + G_i p_i)^T p_{i+1}$$

$$\geq (\nabla f_{i+1} - \nabla f_i - \nabla^2 f_i p_i)^T p_{i+1}$$

$$- [(G_i - \nabla^2 f_i)p_i]^T p_{i+1}$$

(by the minimum principle and $x_i \in \psi$ for large enough i)

By taking norms and using a mean value theorem in [Ortega and Rheinboldt 70, p. 70 Theorem 3.2.5] we have that

$$-\frac{\nu_1}{2}\|p_{i+1}\|^2 \geq (-\eta\|p_i\|^2 - \|(G_i-\nabla^2 f_i)p_i\|)\|p_{i+1}\|$$

Hence

$$0 \leq \frac{\|p_{i+1}\|}{\|p_i\|} \leq \frac{2}{\nu_1}\left(\eta\|p_i\| + \frac{\|(G_i - {}^2 f_i)p_i\|}{\|p_i\|}\right)$$

Since $\{\|p_i\|\} \to 0$ and by assumption the last term in the above inequality converges to zero we finally have that $\lim_{i\to\infty} \frac{\|p_{i+1}\|}{\|p_i\|} = 0$ and by lemma 3.7 we have that $\lim_{i\to\infty} \frac{\|x_{i+1}-\bar{x}\|}{\|x_i-\bar{x}\|} = 0$.

4. Numerical Implementation

The additional condition required on $\{G_i\}$ in theorem 3.3 is satisfied by different updating schemes [Broyden et al 73]. However, to assure the positive definiteness of G_i we use a scheme given in [Garcia-Palomares 73]. For the sake of completeness we briefly describe it, but for brevity we drop the index i and replace i + 1 by 1 :

$$G_1 = G + \frac{\zeta(y-Gs)s^TC}{s^TCs} + \frac{\zeta Cs(y-Gs)^T}{s^TCs} - \frac{\zeta^2(y-Gs)^Ts}{(s^TCs)^2}Css^TC$$

where

$$C_1 = \begin{cases} I & \text{if } i = 0 \bmod n \\ C - \dfrac{\zeta C s s^T C}{s^T C s} \end{cases}$$

and $\zeta \in (0,1)$ is chosen such that G_1 is positive definite. Since in fact $\{G_i\} \to \nabla^2 f(\bar{x})$ we assume additionally that $\nabla^2 f(x)$ is strictly positive definite for all $x \in N(\bar{x},\delta)$ for some $\delta > 0$. To simplify the implementation of the algorithm, we solve the dual of (Q_i) which deals with the updating formulation of the sequence of inverses of G_i, and in this way, $\zeta \in (0,1)$ can be easily computed.

For the stepsize procedure we use the Armijo procedure, which satisfies conditions a and b indicated above (See step 3 and 3' of the algorithm).

The algorithm was tested with problem number one given in [Colville 68] and run on a UNIVAC 1108. The results are in the table on the following page.

RESULTS FOR COLVILLE PROBLEM NUMBER ONE

EXECUTION TIME		.13 secs.
STANDARD TIME*		.00481
MINIMUM FUNCTION VALUE		-32.34868
SOLUTION VECTOR $\bar{x}$:		
	x^1	.300000
	x^2	.333444
	x^3	.400000
	x^4	.428252
	x^5	.224021
BEST STANDARD* TIME IN COLVILLE		.0061

*The standard time is computed, by dividing the execution time of the problem by the execution time (27 seconds for the 1108) of a standard package provided by Colville.

Acknowledgment

I am indebted to Dr. M. J. D. Powell who suggested the proof of lemma 3.7.

References

Armijo, L. (1966) Minimization of Functions Having Lipschitz Continuous First Partial Derivatives. Pacific J. Math., 16 p. 1-3.

Broyden, C. G., Dennis Jr., J. E. and Moré, J. J. (1973) On the Local and Superlinear Convergence of Quasi-Newton Methods. J. Inst. Maths. Appliss. 12, pp. 223-245.

Colville, A. R. (1968) A comparative Study on Nonlinear Programming Codes. IBM New York Scientific Center, Report 320-2949.

Cottle, R. W. and Dantzig, G. B. (1968) "The Principal Pivoting Method of Quadratic Programming" in: Mathematics of the Decision Sciences. Ed. by Dantzig, G. B. and Veinott, A. F. American Mathematics Society, Rhode Island, pp. 244-262.

Daniel, J. W. (1970) "Convergent Step-sizes for Gradient-like Feasible Direction Algorithms for Constrained Optimization" in: Nonlinear Programming. Ed. by Rosen, J. B., Mangasarian, O. L. and Ritter, K. Academic Press, New York pp. 245-274.

Daniel, J. W. (1971) The Approximate Minimization of Functionals. Prentice Hall, New Jersey.

Dennis, J. E. and Moré, J. J. (1974) A Characterization of Superlinear Convergence and Its Application to Quasi-Newton Methods" Math. Comp. 28, pp. 549-560.

García-Palomares, U. M. (1973) Superlinearly Convergent Quasi-Newton Methods for Nonlinear Programming. Ph.D. dissertation, University of Wisconsin, U.S.A.

Mangasarian, O. L. (1969) Nonlinear Programming. McGraw-Hill, New York.

Mangasarian, O. L. (1972) "Dual, Feasible Direction Algorithms" in: Techniques of Optimization. Ed. by Balakrishnan, A. V. Academic Press, New York pp. 67-88.

Ortega, J. M. and Rheinboldt, W. C. (1970) Iterative Solution of Nonlinear Equations of Several Variables. Academic Press, New York.

Ostrowski, A. M. (1966) Solution of Equations and System of Equations. Academic Press, New York.

Pironneau, O. and Polak, E. (1971) Rate of Convergence of a Class of Methods of Feasible Directions, Memo ERL-M301, Univ. of Calif. Berkeley.

Stoer, J. (1971) On the Numerical Solution of Constrained Least Squares Problems. SIAM J. on Num. Anal. 8 pp. 382-411.

Topkis, D. M. and Veinott Jr., A. F. (1967) On the Convergence of Some Feasible Direction Algorithms for Nonlinear Programming. SIAM J. on Control 5 pp. 280-294.

AN IDEAL PENALTY FUNCTION FOR CONSTRAINED OPTIMIZATION[1)]

by

R. Fletcher

The University, Dundee,[2)]

1. Introduction

Powell (1969) has suggested that to solve the problem

$$\begin{array}{llll} \text{minimize} & F(\underset{\sim}{x}) & \underset{\sim}{x} \in R^n, & \\ \text{subject to} & c_i(\underset{\sim}{x}) = 0 & i = 1,2,\ldots,m, & \end{array} \qquad (1.1)$$

in the sense of finding a local minimizer $\underset{\sim}{x}^*$, a suitable penalty function is

$$\begin{aligned} \phi(\underset{\sim}{x},\underset{\sim}{\theta},S) &= F(\underset{\sim}{x}) + \tfrac{1}{2}(\underset{\sim}{c}(\underset{\sim}{x})-\underset{\sim}{\theta})^T S(\underset{\sim}{c}(\underset{\sim}{x})-\underset{\sim}{\theta}) \\ &= F(\underset{\sim}{x}) + \tfrac{1}{2}\sum_i \sigma_i(c_i(\underset{\sim}{x})-\theta_i)^2 \end{aligned} \qquad (1.2)$$

where $\underset{\sim}{\theta} \in R^m$, and S is an $m \times m$ diagonal matrix with diagonal elements $\sigma_i > 0$. (In this presentation the signs of the θ_i have been changed from those used by Powell, and a factor ½ introduced in (1.2) to simplify the later analysis.) The penalty function is used in the usual way, that is for any given value of

---

1) This paper has been published in the Journal of the Institute of Mathematics and Its Applications. It is being published here at the invitation of the editors and with the permission of the Institute of Mathematics and Its Applications. There are some small differences which are explained in the 'Note in Proof.'

2) Much of this work was carried out whilst the author was at AERE Harwell.

the parameters $\underset{\sim}{\theta}$,S, a vector $\underset{\sim}{x}(\underset{\sim}{\theta},S)$ is obtained which minimizes $\phi(\underset{\sim}{x},\underset{\sim}{\theta},S)$ without constraints. There is an outer iteration in which $\underset{\sim}{\theta}$ and S are changed so as to cause the solutions $\underset{\sim}{x}(\underset{\sim}{\theta},S) \rightarrow \underset{\sim}{x}^*$. A well known peanlty function is one with $\underset{\sim}{\theta} = 0$, in which case this convergence is ensured by letting $\sigma_i \rightarrow \infty$, $i=1,2,...,m$. However Powell suggests an outer iteration for use with (1.2) such that it is not necessary to force $\sigma_i \rightarrow \infty$ in order to achieve convergence. Rather the aim is to keep S constant and to let $\underset{\sim}{\theta} \rightarrow \underset{\sim}{\theta}^*$, where $\underset{\sim}{\theta}^*$ is an optimum vector of parameters satisfying

$$\theta_i^* \, \sigma_i = \lambda_i^* \qquad i = 1,2,...,m \tag{1.3}$$

where $\underset{\sim}{\lambda}^*$ is the vector of Lagrange multipliers for the solution $\underset{\sim}{x}^*$ to (1.1). It is only necessary to increase the σ_i when the rate of convergence of $\underset{\sim}{x}(\underset{\sim}{\theta},S)$ to $\underset{\sim}{x}^*$ is not sufficiently rapid. The method is explained in more detail in sections 3 and 4.

At about the same time, and independently of Powell, Hestenes (1969) put forward what he called the method of multipliers. In this he suggested using the penalty function

$$\psi(\underset{\sim}{x},\underset{\sim}{\lambda},S) = F(\underset{\sim}{x}) - \underset{\sim}{\lambda}^T\underset{\sim}{c}(\underset{\sim}{x}) + \tfrac{1}{2}\underset{\sim}{c}(\underset{\sim}{x})^T S \underset{\sim}{c}(\underset{\sim}{x}) \tag{1.4}$$

where $\underset{\sim}{\lambda} \in R^m$ and S is as above. (In fact Hestenes uses $S = \sigma I$ and therefore implicitly assumes that the constraints are well scaled.) If (1.4) is minimized for fixed $\underset{\sim}{\lambda}$,S, then a vector $\underset{\sim}{x}(\underset{\sim}{\lambda},S)$ is obtained. It is clear on expanding (1.2) that if

$$\theta_i \, \sigma_i = \lambda_i \qquad i = 1,2,...,m, \tag{1.5}$$

then

$$\phi(\underset{\sim}{x},\underset{\sim}{\theta},S) = \psi(\underset{\sim}{x},\underset{\sim}{\lambda},S) + \tfrac{1}{2}\Sigma \, \lambda_i^2/\sigma_i \; . \tag{1.6}$$

Because the difference between ϕ and ψ is independent of $\underset{\sim}{x}$, it follows that $\underset{\sim}{x}(\underset{\sim}{\lambda},S) = \underset{\sim}{x}(\underset{\sim}{\theta},S)$ for any S, if $\underset{\sim}{\lambda}$ and $\underset{\sim}{\theta}$ are related by (1.5). However the penalty function values $\phi(\underset{\sim}{x}(\underset{\sim}{\theta},S),\underset{\sim}{\theta},S)$ and $\psi(\underset{\sim}{x}(\underset{\sim}{\lambda},S),\underset{\sim}{\lambda},S)$ differ, and this difference turns out to be important. Given these relationships between $\underset{\sim}{\theta}$ and $\underset{\sim}{\lambda}$ the iterative methods suggested by Powell and by Hestenes for changing the $\underset{\sim}{\theta}$ (or $\underset{\sim}{\lambda}$) parameters are the same. However Powell goes into the situation in much more detail and also suggests an algorithm for increasing S which enables him to prove strong convergence results.

The work in this paper was originally motivated by attempting to modify Powell's function (1.2) to solve the inequality problem

$$\begin{aligned} &\text{minimize} \quad F(\underset{\sim}{x}) \\ &\text{subject to} \quad c_i(\underset{\sim}{x}) \geq 0, \quad i = 1,2,\ldots,m. \end{aligned} \tag{1.7}$$

by using the penalty function

$$\Phi(\underset{\sim}{x},\underset{\sim}{\theta},S) = F(\underset{\sim}{x}) + \tfrac{1}{2}\sum_i \sigma_i(c_i(\underset{\sim}{x})-\theta_i)_-^2, \tag{1.8}$$

where

$$a_- = \min(a,0) = \begin{cases} a & \text{if} \quad a \leq 0 \\ 0 & \text{if} \quad a \geq 0. \end{cases}$$

In fact there is no difficulty in generalizing (1.8) further to deal with problems with mixed equality/inequality constraints, but the aim here is to keep the notation simple. As before, the case $\underset{\sim}{\theta} = 0$ is well known, and convergence can be forced by letting $\sigma_i \to \infty$ $i = 1,2,\ldots,m$. However difficulties then arise because the second derivative jump discontinuities in (1.8) tend to infinity, and also occur at points which tend to $\underset{\sim}{x}^*$. The effect of using the $\underset{\sim}{\theta}$ parameters of (1.8) to solve an inequality problem can be illustrated simply. Consider the one variable problem: minimize $F(x)$ subject to $c(x) \geq 0$. If an initial choice $\theta = 0$, $\sigma = 1$ is made (assuming the latter is sufficiently large), then the penalty term is only effective for $c < 0$ and the minimum of $\Phi(x,0,1)$ is at $c(x) = c_{min} < 0$ (see figure 1). If the correction $\theta' = \theta - c_{min}$ is made, (as suggested by Powell and Hestenes), then for the function $\Phi(x,\theta',1)$, the penalty term is effective when $c < \theta'$, and a minimum of Φ is created in the neighbourhood of the solution at $c(x) = 0$.

In this paper it will be assumed that $F(\underset{\sim}{x})$ and $c_i(\underset{\sim}{x})$ $i = 1,2,\ldots,m$, are twice continuously differentiable. Under these circumstances, $\Phi(\underset{\sim}{x})$ is also twice continuously differentiable except at points $\underset{\sim}{x}$ for which any $c_i(\underset{\sim}{x}) = \theta_i$, where the second derivative has a jump discontinuity. However the size of this discontinuity is bounded above when S is bounded, and usually is remote from the minimum, as in figure 1, where it does not much affect convergence of the minimization routine.

In fact a function closely related to (1.8) has already been suggested by Rockafellar, originally in spoken form at the 7th International Mathematical Programming Symposium at The Hague (1970), and more recently in manuscript form (Rockafellar, 1973a,b, 1974). The idea is to modify the Hestenes function (1.4) giving

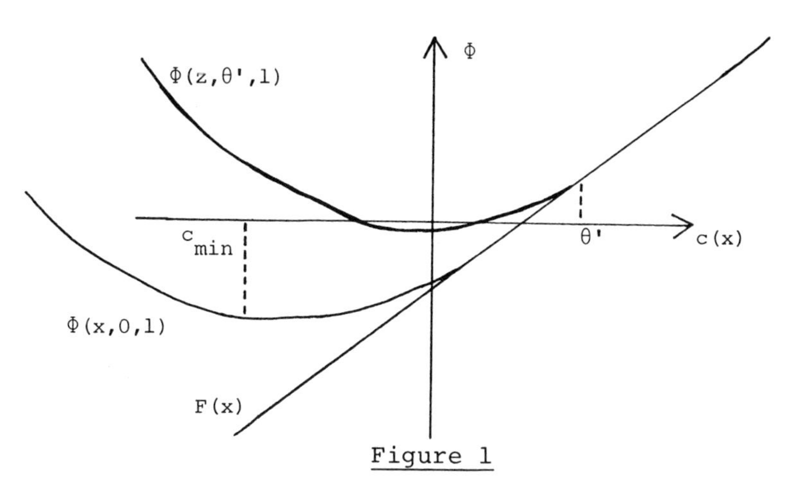

Figure 1

The penalty function (1.8) for an inequality problem

$$\Psi(\underset{\sim}{x},\underset{\sim}{\lambda},S) = F(\underset{\sim}{x}) + \sum_i \begin{cases} -\lambda_i c_i + \frac{1}{2}\sigma_i c_i^2 & \text{if } c_i \le \lambda_i/\sigma_i \\ -\frac{1}{2}\lambda_i^2/\sigma_i & \text{if } c_i \ge \lambda_i/\sigma_i \end{cases} \tag{1.9}$$

Actually Rockafellar also considers the more simple case $\sigma_i = \sigma$ $i = 1,2,\ldots,m$. It is easy to see that the same relationship to (1.6) holds between Ψ and Φ, namely

$$\Phi(\underset{\sim}{x},\underset{\sim}{\theta},S) = \Psi(\underset{\sim}{x},\underset{\sim}{\lambda},S) + \tfrac{1}{2}\sum_i \lambda_i^2/\sigma_i, \tag{1.10}$$

and that if $\underset{\sim}{x}(\underset{\sim}{\lambda},S)$ minimizes (1.9), then $\underset{\sim}{x}(\underset{\sim}{\lambda},S) = \underset{\sim}{x}(\underset{\sim}{\theta},S)$ where $\underset{\sim}{x}(\underset{\sim}{\theta},S)$ minimizes (1.8), assuming (1.5). Rockafellar has been concerned primarily with the structure of the function $\Psi(\underset{\sim}{x}(\underset{\sim}{\lambda},S),\underset{\sim}{\lambda},S)$ because strong duality results can be derived when the original problem (1.7) satisfies certain convexity assumptions. Arrow et al (1973) have also considered the idea and they give local duality results, but their results are unnecessarily restrictive. In his

Ph.D. thesis, Buys (1972) is also interested in local results proved without convexity assumptions. Mangasarian (1973) has also investigated developments of the same idea in various directions and there is now much interest in the field, evident by the number of abstracts submitted to the recent 8th Mathematical Programming Symposium at Stanford.

Another antecedent to (1.8) is the barrier function

$$\Phi(\underset{\sim}{x},\underset{\sim}{\theta},S) = F(\underset{\sim}{x}) - \sum_i \sigma_i \log_r(c_i(\underset{\sim}{x})-\theta_i)$$

suggested by M. Osborne (private communication), in which the idea is to introduce $\underset{\sim}{\theta}$ parameters into other well known barrier functions. Clearly (1.8) is a variation of this same idea.

In section 2 of this paper a review of the local duality results is given, with particular emphasis on those of practical importance. Buys' (1972) development for the equality problem is largely followed, but in stating the corresponding results for the inequality problem (without proof), Busy seems to be mistaken about the result given here in (2.15) and which is of much importance. The result given here in Theorem 4 is also thought to be new. In section 3 of this paper it is shown what implications the results of section 2 have for Newton like methods to adjust the $\underset{\sim}{\lambda}$ parameters. Again Powell (1969), Buys (1972) and Mangasarian (1973) are all aware of this possibility, but it is shown here that such methods can usefully be used even when second derivatives are not available, by minimizing $\phi(\underset{\sim}{x},\underset{\sim}{\theta},S)$ by a quasi-Newton method, and extracting the approximate hessian for use in the λ iteration. Buys is the only person who has considered computational problems for inequality constraints, but because of his apparent mistake about (2.15), he recommends avoiding the Newton-like iteration in this case. In fact this paper indicates theoretically that the iteration is eminently suitable and this is strongly backed up by computational experience. Another way of iterating in the λ parameters is that due to Powell and to Hestenes, and a simple derivation is given of its relation to the Newton iteration.

The convergence of the iterations described in section 3 is only local, and the iteration must be supplemented by a strategy for increasing S so as to force convergence. The strategy due to Powell (1969) is described, and it is pointed out that it is readily adapted to the inequality problem. Powell's strategy varies either λ or S on an iteration, but not both. In fact it is shown that it is consistent to consider algorithms which vary both λ and S at every iteration. Three possibilities are suggested. In section 5 the results of extensive numerical tests are described. All the variations are shown to perform well and their individual merits are discussed. The best variation is compared against other penalty and barrier functions, and it is argued that it is superior from both theoretical and practical considerations.

As to notation, the operators ∇ and ∇^2 will refer to $[\partial/\partial x_i]$ and $[\partial^2/(\partial x_i \partial x_j)]$ respectively unless specifically qualified (as ∇_λ). The definitions $g = \nabla\phi = \nabla\psi$ and $G = \nabla^2\phi = \nabla^2\psi$ will be used. For the equality problem (1.1), $N(\underset{\sim}{x})$ will refer to the matrix $[\underset{\sim}{\nabla} c_1, \underset{\sim}{\nabla} c_2, \ldots, \underset{\sim}{\nabla} c_m]_{\underset{\sim}{x}}$ and it will be assumed that $N(\underset{\sim}{x}^*) = N^*$ is of full rank. In this case there exist unique multipliers $\underset{\sim}{\lambda}^*$ such that

$$\underset{\sim}{\nabla} F^* = N^* \underset{\sim}{\lambda}^* \; . \tag{1.11}$$

For the inequality problem, if $\underset{\sim}{x}^*$ is a strong local minimizer of (1.7) the set of <u>weakly active</u> constraints $\bar{A}^* = \{i: c_i(\underset{\sim}{x}^*)=0\}$ can be defined. It will be assumed that the vectors $\underset{\sim}{\nabla} c_i \;\; i \varepsilon \bar{A}^*$ are linearly independent, in which case there exist unique multipliers $\underset{\sim}{\lambda}^*$ such that

$$\lambda_i^* \geq 0 \quad i \;\varepsilon\; \bar{A}^*, \;\; \lambda_i^* = 0 \quad i \not\varepsilon\; \bar{A}^* \tag{1.12a}$$

$$\underset{\sim}{\nabla} F^* = \sum_{i \varepsilon A^*} \underset{\sim}{\nabla} c_i(\underset{\sim}{x}^*) \; \underset{\sim}{\lambda}_i^* \tag{1.12b}$$

where $A^* = \{i: \lambda_i^* > 0\}$ is the set of <u>strongly active</u> constraints. Note that $\bar{A}^* \supseteq A^*$.

2. <u>Optimality results for Lagrange multipliers</u>

In this section some important duality results will be developed, showing that the optimum choice of the $\underset{\sim}{\lambda}$ (or $\underset{\sim}{\theta}$) parameters for the Powell/Hestenes/ Rockafellar penalty function is determined by a maximization problem in terms of these parameters. This problem is unconstrained, even for the inequality penalty function. First of all the equality problem (1.1) is considered, subject to the various assumptions of section 1. Initially a thoerem is proved which shows that if the optimum multipliers $\underset{\sim}{\lambda}^*$ are chosen in constructing $\psi(\underset{\sim}{x}, \underset{\sim}{\lambda}^*, S)$ or $\phi(\underset{\sim}{x}, \underset{\sim}{\theta}^*, S)$ where $\underset{\sim}{\lambda}^* = S\underset{\sim}{\theta}^*$, then $\underset{\sim}{x}^*$ is a strong local minimum of both these functions, and hence $\underset{\sim}{x}^*$ is $\underset{\sim}{x}(\underset{\sim}{\lambda}^*)$.

Theorem 1 If the second order conditions (2.3 below) on the problem are satisfied, then there exists an $S' > 0$ such that for any $S \geq S'$, $\underset{\sim}{x}^*$ is a strong local minimum with respect to $\underset{\sim}{x}$ of both $\phi(\underset{\sim}{x},\underset{\sim}{\theta}^*,S)$ and $\psi(\underset{\sim}{x},\underset{\sim}{\lambda}^*,S)$.

Proof The argument is first to show that the necessary conditions $\underset{\sim}{c}(\underset{\sim}{x}^*) = 0$ and (1.11) imply that $\underset{\sim}{\nabla}\phi(\underset{\sim}{x}^*,\underset{\sim}{\theta}^*,S) = 0$ which is a necessary condition for $\underset{\sim}{x}^*$ to be a local minimum of ϕ. This result follows directly from the equation

$$\underset{\sim}{\nabla}\phi(\underset{\sim}{x},\underset{\sim}{\theta},S) = \underset{\sim}{\nabla}F(\underset{\sim}{x}) + NS(\underset{\sim}{c}(\underset{\sim}{x})-\underset{\sim}{\theta}). \tag{2.1}$$

Given this result, then a sufficient condition for the theorem to hold is to show that $\nabla^2\phi(\underset{\sim}{x}^*,\underset{\sim}{\theta}^*,S)$ is positive definite. Now from (2.1)

$$\nabla^2\phi(\underset{\sim}{x}^*,\underset{\sim}{\theta}^*,S) = L^* + N^*SN^{*T}$$

where

$$L^* = \nabla^2F(\underset{\sim}{x}^*) - \sum_i \lambda_i^*\nabla^2 c_i(\underset{\sim}{x}^*).$$

Consider any vector $\underset{\sim}{u}$ such that $||\underset{\sim}{u}||_2 = 1$ and let $\underset{\sim}{u}$ be written as a component $\underset{\sim}{v}$ orthogonal to the columns of N^*, plus a component which is a linear combination of the columns of N^*. This can be conveniently written

$$\underset{\sim}{u} = \underset{\sim}{v} + N^{*+T}\underset{\sim}{w}$$

where $A^+ = (A^TA)^{-1}A^T$. Then

$$\underset{\sim}{u}^T\nabla^2\phi\underset{\sim}{u} = \underset{\sim}{v}^TL^*\underset{\sim}{v} + 2\underset{\sim}{v}^TL^*N^{*+T}\underset{\sim}{w} + \underset{\sim}{w}^TN^{*+}L^*N^{*+T}\underset{\sim}{w} + \underset{\sim}{w}^TS\underset{\sim}{w} \tag{2.2}$$

Now sufficient conditions for a solution of (1.1) are that there exists an $a > 0$ such that

$$v^T L^* v \geq a \|v\|_2^2 \quad \forall\, v:\ N^{*T} v = 0. \tag{2.3}$$

Writing $\|L^* N^{*+T}\|_2 = b$ and $\|N^{*+} L^* N^{*+T}\|_2 = d$, then from (2.2)

$$u^T \nabla^2 \phi u \quad a\|v\|_2^2 - 2b\|v\|_2 \|w\|_2 + (\min_i \sigma_i - d)\|w\|_2^2.$$

Because $v = w = 0$ cannot hold, if $S' = \sigma' I$ is chosen so that $\sigma' > d + b^2/a$, then for all $S \geq S'$ it follows that $u^T \nabla^2 \phi u > 0$. This establishes the positive definiteness of $\nabla^2 \phi$ and the theorem follows in respect of ϕ. The same result holds for $\psi(x, \lambda^*, S)$ because (1.6) shows that the difference between ϕ and ψ is independent of x. ∎

It is instructive that it is not worthwhile attempting to strengthen this theorem, because when the conditions (2.3) do not hold, then it may not be possible to solve (1.1) by minimizing the function $\phi(x, \theta^*, S)$ for any S. A simple example due to Hestenes (private communication) illustrates the point well. Let $F(x) = x_1^4 + x_1 x_2$ and $c(x) = x_2$. Then although $x^* = 0$ solves (1.1) with optimum multiplier $\lambda^* = 0$, there is no value of σ for which $\phi(x, 0, S) = x_1^4 + x_1 x_2 + \frac{1}{2} x_2^2$ is minimized at x^*. Henceforth it will be assumed that the conditions (2.3) hold.

For much of this paper it will be the case that a fixed value of S has been chosen which satisfies Thoerem 1, and that the interest is in determining optimum λ (or θ) parameters. In these cases the explicit dependence of $\phi, \psi, \underset{\sim}{x}$, etc. on S can and will be dropped. The subsequent analysis can be carried out equivalently either in terms of the parameters $\underset{\sim}{\theta}$ or the parameters $\underset{\sim}{\lambda} = S\underset{\sim}{\theta}$. However it turns out to be convenient to work in terms of the $\underset{\sim}{\lambda}$ parameters for the most part.

It is useful to regard $\underset{\sim}{x}(\underset{\sim}{\lambda})$ as a function implicitly determined by solving the nonlinear equations

$$\underset{\sim}{\nabla}\psi(\underset{\sim}{x}, \underset{\sim}{\lambda}) = 0. \tag{2.4}$$

Because $\nabla^2\psi(\underset{\sim}{x}^*, \underset{\sim}{\lambda}^*)$ is positive definite, it follows from the implicit function theorem (Hestenes, 1966, for example) that there exists an open neighbourhood $\Omega \subset R^m$ about $\underset{\sim}{\lambda}^*$ and a function $\underset{\sim}{x}(\underset{\sim}{\lambda})$ defined on Ω such that both $\underset{\sim}{x}(\underset{\sim}{\lambda})$ is continuous and continuously differentiable and that $\nabla^2\psi(\underset{\sim}{x}(\underset{\sim}{\lambda}), \underset{\sim}{\lambda})$ is positive definite for all $\underset{\sim}{\lambda} \in \Omega$. It may be of course that $\psi(\underset{\sim}{x}, \underset{\sim}{\lambda})$ has local minima, so that various solutions to (2.4) exist. It is assumed however that a consistent choice of $\underset{\sim}{x}(\underset{\sim}{\lambda})$ is made by the minimization routine, that is the particular solution which exists by virtue of the implicit function theorem.

It is important to look at other quantities derived from $\underset{\sim}{x}(\underset{\sim}{\lambda})$ for any given $\underset{\sim}{\lambda}$, in particular $\underset{\sim}{c}(\underset{\sim}{x}(\underset{\sim}{\lambda}))$ and $\psi(\underset{\sim}{x}(\underset{\sim}{\lambda}), \underset{\sim}{\lambda})$. For convenience these will be written $\underset{\sim}{c}(\underset{\sim}{\lambda})$ and $\psi(\underset{\sim}{\lambda})$, with the convention that if the dependence on $\underset{\sim}{x}$ is not written explicitly, then $\underset{\sim}{x}(\underset{\sim}{\lambda})$ is implied. Because $\underset{\sim}{x}(\underset{\sim}{\lambda})$ is differentiable, it is possible to develop an expression for the Jacobian matrix $[d\underset{\sim}{c}/d\underset{\sim}{\lambda}]$ of $\underset{\sim}{c}(\underset{\sim}{\lambda})$, where $[dc/d\lambda]_{ij} = dc_i/d\lambda_j$ and where the total differential is used in the sense that $\underset{\sim}{x}$ is not held constant but is the function $\underset{\sim}{x}(\underset{\sim}{\lambda})$. Because $\underset{\sim}{c}(\underset{\sim}{\lambda}) = \underset{\sim}{c}(\underset{\sim}{x}(\underset{\sim}{\lambda}))$, the chain rule implies that

$$[d\underset{\sim}{c}/d\underset{\sim}{\lambda}] = [\partial\underset{\sim}{c}/\partial\underset{\sim}{x}]\,[\partial\underset{\sim}{x}/\partial\underset{\sim}{\lambda}] = N^T[\partial\underset{\sim}{x}/\partial\underset{\sim}{\lambda}]. \qquad (2.5)$$

Operating on (2.4) by $[d/d\underset{\sim}{\lambda}]$ gives

$$[d\underset{\sim}{\nabla}\psi(\underset{\sim}{x}(\underset{\sim}{\lambda}),\underset{\sim}{\lambda})/d\underset{\sim}{\lambda}] = [\partial\underset{\sim}{\nabla}\psi/\partial\underset{\sim}{x}]\,[\partial\underset{\sim}{x}/\partial\underset{\sim}{\lambda}] + [\partial\underset{\sim}{\nabla}\psi/\partial\underset{\sim}{\lambda}] = 0. \qquad (2.6)$$

But $[\partial\underset{\sim}{\nabla}\psi/\partial\underset{\sim}{x}] = \nabla^2\psi(\underset{\sim}{x}(\underset{\sim}{\lambda}),\underset{\sim}{\lambda}) = G(\underset{\sim}{\lambda})$ say, and $[\partial\underset{\sim}{\nabla}\psi/\partial\lambda] = -N$, so from (2.5) and (2.6) there follows

$$[d\underset{\sim}{c}/d\underset{\sim}{\lambda}] = [N^TG^{-1}N]_{\underset{\sim}{x}(\underset{\sim}{\lambda})} \qquad (2.7)$$

This result is significant in that it shows that the Jacobian matrix of $\underset{\sim}{c}(\underset{\sim}{\lambda})$ is symmetric positive definite at $\underset{\sim}{\lambda}^*$ and positive semi-definite for all $\underset{\sim}{\lambda} \in \Omega$. Therefore it can be expected that there exists a (convex) function of λ defined on Ω whose gradient is $\underset{\sim}{c}(\underset{\sim}{\lambda})$ and whose hessian matrix is $N^TG^{-1}N$. In fact Buys (1972) points out that $-\psi(\underset{\sim}{\lambda})$ is this particular function. This follows because

$$[d\psi/d\underset{\sim}{\lambda}] = [\partial\psi/\partial\underset{\sim}{x}]\,[\partial\underset{\sim}{x}/\partial\underset{\sim}{\lambda}] + [\partial\psi/\partial\underset{\sim}{\lambda}]$$

$$= -\,\underset{\sim}{c}(\underset{\sim}{\lambda})$$

by virtue of equation (2.4) and by $\partial\psi/\partial\lambda_i = -c_i$ from (1.4). Thus $\psi(\underset{\sim}{\lambda})$ has derivatives

$$\underset{\sim}{\nabla}_\lambda\psi(\underset{\sim}{\lambda}) = -\underset{\sim}{c}(\underset{\sim}{\lambda}) \qquad (2.8a)$$

and

$$\nabla^2_\lambda\psi(\underset{\sim}{\lambda}) = -\,[N^TG^{-1}N]_{\underset{\sim}{\lambda}}\,. \qquad (2.8b)$$

The main optimality result for the equality problem can now be derived. Theorem 1 implies that

$\underset{\sim}{x}(\underset{\sim}{\lambda}^*)$ solves (1.1) and so is $\underset{\sim}{x}^*$. Therefore $\underset{\sim}{c}(\underset{\sim}{\lambda}^*) = \underset{\sim}{c}(\underset{\sim}{x}(\underset{\sim}{\lambda}^*)) = 0$ and hence $\underset{\sim}{\lambda}^*$ is a stationary point of $\psi(\underset{\sim}{\lambda})$ by (2.8a). Because $\nabla_\lambda^2 \psi(\underset{\sim}{\lambda}^*)$ is negative definite and because $\psi(\underset{\sim}{\lambda})$ is concave, $\underset{\sim}{\lambda}^*$ is the only maximizer of $\psi(\underset{\sim}{\lambda})$ on Ω and hence is a strong local maximizer of $\psi(\underset{\sim}{\lambda})$ on R^m. Thus the problem of finding optimum Lagrange multipliers has been formulated in terms of an optimization problem defined in terms of $\psi(\underset{\sim}{\lambda})$. The problem of minimizing the penalty function $\psi(\underset{\sim}{x},\underset{\sim}{\lambda})$ with respect to $\underset{\sim}{x}$ is embedded within the problem of finding a maximizer of $\psi(\underset{\sim}{\lambda})$ with respect to $\underset{\sim}{\lambda}$. A study of various iterative methods for solving this latter problem will be made in section 3. If the functions $F(\underset{\sim}{x})$ and $c_i(\underset{\sim}{x})$ have special features which imply $\underset{\sim}{x}(\underset{\sim}{\lambda})$ to be the global minimizer of $\psi(\underset{\sim}{x},\underset{\sim}{\lambda})$, then $\underset{\sim}{\lambda}^*$ is the global maximizer of $\psi(\underset{\sim}{\lambda})$. This follows from the inequality

$$\psi(\underset{\sim}{x}(\underset{\sim}{\lambda}),\underset{\sim}{\lambda}) \le \psi(\underset{\sim}{x}^*,\underset{\sim}{\lambda}) = \psi(\underset{\sim}{x}^*,\underset{\sim}{\lambda}^*) \tag{2.9}$$

by virtue of $\underset{\sim}{c}(\underset{\sim}{x}^*) = 0$, a result first used by Powell (1969).

These results concerning the optimality of the multipliers $\underset{\sim}{\lambda}^*$ can all be extended to cover the inequality problem (1.7) by using the penalty function $\Psi(\underset{\sim}{x},\underset{\sim}{\lambda})$ defined in (1.9). As before S will be fixed and $\underset{\sim}{\lambda}$ or $\underset{\sim}{\theta}$ used interchangeably subject to $\underset{\sim}{\lambda} = S\underset{\sim}{\theta}$. In order to prove a result analogous to Theorem 1, it is useful to state sufficient conditions for the function $\Phi(\underset{\sim}{x},\underset{\sim}{\theta},S)$ defined in (1.8) to have a strong local minimum. The usual conditions break down when $\nabla^2\Phi$ exhibits a jump discontinuity at $\underset{\sim}{x}(\underset{\sim}{\theta})$. Let $\underset{\sim}{x}(\underset{\sim}{\lambda})$ minimize $\Psi(\underset{\sim}{x},\underset{\sim}{\lambda})$ and hence $\Phi(\underset{\sim}{x},\underset{\sim}{\theta})$ with respect to $\underset{\sim}{x}$, in which case the index set of constraints which contribute to Φ can be defined, namely

$$M(\underset{\sim}{\lambda}) = \{i: c_i(\underset{\sim}{x}(\underset{\sim}{\lambda}))<\theta_i\}. \tag{2.10}$$

Then a sufficient condition for the function

$$\phi'(\underset{\sim}{x},\underset{\sim}{\theta}) = F(\underset{\sim}{x}) + \tfrac{1}{2} \sum_{i\varepsilon M(\underset{\sim}{\lambda})} \sigma_i(c_i(\underset{\sim}{x})-\theta_i)^2 \tag{2.11}$$

to have a strong local minimum at $\underset{\sim}{x}(\underset{\sim}{\lambda})$ is that $\nabla^2\phi'(\underset{\sim}{x}(\underset{\sim}{\lambda}),\underset{\sim}{\theta})$ is positive definite (assuming that $\underset{\sim}{\nabla}\phi = 0$). But $\Phi(\underset{\sim}{x},\underset{\sim}{\theta})$ satisfies $\Phi(\underset{\sim}{x}(\underset{\sim}{\lambda}),\underset{\sim}{\theta}) = \phi'(\underset{\sim}{x}(\underset{\sim}{\lambda}),\underset{\sim}{\theta})$, and $\Phi(\underset{\sim}{x},\underset{\sim}{\theta}) \geq \phi'(\underset{\sim}{x},\underset{\sim}{\theta})$ for $\underset{\sim}{x}$ in some neighbourhood of $\underset{\sim}{x}(\underset{\sim}{\lambda})$. Hence the condition on $\nabla^2\phi'$ is sufficient to ensure a strong local minimum of $\Phi(\underset{\sim}{x},\underset{\sim}{\theta})$. A result analogous to Theorem 1 can now be stated and proved.

<u>Theorem 2</u> If second order sufficiency conditions (2.12) below are satisfied by the inequality problem (1.7), then there exists an $S' > 0$ such that for all $S \geq S'$, $\underset{\sim}{x}^*$ is a strong local minimum of $\Phi(\underset{\sim}{x},\underset{\sim}{\theta}^*,S)$ with respect to $\underset{\sim}{x}$, where $\underset{\sim}{\lambda}^* = S\underset{\sim}{\theta}^*$.

<u>Proof</u> The necessary conditions $\underset{\sim}{\nabla}\Phi = 0$ follow from (1.12b) in the same way as for Theorem 1. Not very restrictive sufficiency conditions for the problem (1.7) to have a strong local minimum at $\underset{\sim}{x}^*$ are to assume that there exists an $a > 0$ such that

$$\underset{\sim}{v}^T L^* \underset{\sim}{v} \geq a \,||\underset{\sim}{v}||_2^2 \quad \forall\ \underset{\sim}{v}:\ \underset{\sim}{v}^T \underset{\sim}{\nabla} c_i^* = 0 \quad \forall\ i\varepsilon A^*. \tag{2.12}$$

(see Fiacco and McCormick, 1968). But $A^* = M(\underset{\sim}{\lambda}^*)$ so it follows by the proof of Theorem 1 that $\nabla^2\phi'$ is positive definite. It has already been remarked that this is a sufficient condition for a strong local minimum of $\Phi(\underset{\sim}{x},\underset{\sim}{\theta}^*,S)$. As in Theorem 1, an identical result holds for $\Psi(\underset{\sim}{x},\underset{\sim}{\lambda}^*,S)$. ∎

As before the method may not work if the sufficient conditions (2.12) do not hold, so henceforth they will be assumed to hold. Also $\underset{\sim}{x}(\underset{\sim}{\lambda})$ will be deemed to be defined implicitly by the solution of the equations

$$\underset{\sim}{\nabla}\Psi(\underset{\sim}{x},\underset{\sim}{\lambda}) = 0 \qquad (2.13)$$

for $\underset{\sim}{x}$, and hence by the implicit function theorem there exists a neighbourhood $\Omega \subset R^m$ about $\underset{\sim}{\lambda}^*$ in which $\underset{\sim}{x}(\underset{\sim}{\lambda})$ is continuous and $\nabla^2\phi'(\underset{\sim}{x}(\underset{\sim}{\lambda}),\underset{\sim}{\theta})$ is positive definite. Because $\underset{\sim}{\nabla}\psi(\underset{\sim}{x},\underset{\sim}{\lambda})$ is not differentiable, the implicit function theorem states that $\underset{\sim}{x}(\underset{\sim}{\lambda})$ is not differentiable. However because of the identity between $\phi'(\underset{\sim}{x}(\underset{\sim}{\lambda}),\underset{\sim}{\theta},\underset{\sim}{S})$ and $\Phi(\underset{\sim}{x}(\underset{\sim}{\lambda}),\underset{\sim}{\theta},S)$, $\underset{\sim}{x}(\underset{\sim}{\lambda})$ is differentiable except for values of $\underset{\sim}{\lambda}$ for which at least one equation of the form

$$c_i(\underset{\sim}{x}(\underset{\sim}{\lambda})) = \theta_i \quad (=\lambda_i/\sigma_i)$$

is satisfied. These equations form surfaces in $\underset{\sim}{\lambda}$-space on which the derivative of $\underset{\sim}{x}(\underset{\sim}{\lambda})$ has a jump discontinuity.

Again it is convenient to write $\Psi(\underset{\sim}{x}(\underset{\sim}{\lambda}),\underset{\sim}{\lambda})$ as $\Psi(\underset{\sim}{\lambda})$ and $\underset{\sim}{c}(\underset{\sim}{x}(\underset{\sim}{\lambda}))$ as $\underset{\sim}{c}(\underset{\sim}{\lambda})$. Following the analysis for the equality problem, it is interesting to examine derivatives of $\Psi(\underset{\sim}{\lambda})$. The first derivative is given by

$$[d\Psi/d\underset{\sim}{\lambda}] = [\partial\Psi/\partial\underset{\sim}{x}]\,[\partial\underset{\sim}{x}/\partial\underset{\sim}{\lambda}] + [\partial\Psi/\partial\underset{\sim}{\lambda}].$$

Because $\underset{\sim}{x}(\underset{\sim}{\lambda})$ satisfies (2.13), so

$$d\Psi(\underset{\sim}{\lambda})/d\lambda_i = \partial\Psi/\partial\lambda_i = \begin{cases} -c_i & \text{if} \quad c_i < \theta_i \\ -\theta_i & \text{if} \quad c_i \geq \theta_i \end{cases}$$

$$= -\min(c_i,\theta_1), \qquad (2.14)$$

a result first pointed out by Rockafellar. A further differentiation shows that

$$\nabla_\lambda^2 \Psi = \begin{array}{c} i \in M(\underset{\sim}{\lambda}) \\ i \notin M(\underset{\sim}{\lambda}) \end{array} \left\{ \begin{array}{|c|c|} \hline -N^T G^{-1} N & 0 \\ \hline 0 & -S^{-1} \\ \hline \end{array} \right. \qquad (2.15)$$

where the columns of N are taken from the constraints $i \in M(\underset{\sim}{\lambda})$ and those of S^{-1} from the constraints $i \notin M(\underset{\sim}{\lambda})$. Because $\nabla_\lambda^2 \Psi$ is negative semi-definite, $\Psi(\underset{\sim}{\lambda})$ is concave on Ω.

These results can be brought together to demonstrate the optimality properties of Lagrange multipliers in the inequality problem. By Theorem 2, $\underset{\sim}{x}(\underset{\sim}{\lambda}^*)$ solves the problem (1.7) and hence from the necessary conditions (1.12a) and the definition of $\bar{A}^*$ it follows that $\min(c_i^*, \lambda_i^*/\sigma_i) = 0$ for all i. Therefore by (2.14), $\underset{\sim}{\lambda}^*$ is a stationary point of $\Psi(\underset{\sim}{\lambda})$ and in fact a strong local maximizer by the concavity of $\Psi(\underset{\sim}{\lambda})$ and the negative definiteness of $\nabla_\lambda^2 \Psi(\underset{\sim}{\lambda}^*)$. Again various iterative methods for maximizing $\Psi(\underset{\sim}{\lambda})$ will be discussed in section 3. A stronger result, analogous to (2.9) can be stated when the problem (1.7) is such that $\underset{\sim}{x}(\underset{\sim}{\lambda})$ can be guaranteed to be the <u>global</u> minimizer of $\Psi(\underset{\sim}{x}, \underset{\sim}{\lambda})$. Then the inequality

$$\Psi(\underset{\sim}{x}(\underset{\sim}{\lambda}), \underset{\sim}{\lambda}) \leq \Psi(\underset{\sim}{x}^*, \underset{\sim}{\lambda}) \leq \Psi(\underset{\sim}{x}^*, \underset{\sim}{\lambda}) + \tfrac{1}{2} \sum_{i:\lambda_i<0} \lambda_i^2/\sigma_i = \Psi(\underset{\sim}{x}^*, \underset{\sim}{\lambda}^*)$$

holds, showing that $\underset{\sim}{\lambda}^*$ is a <u>global</u> maximizer of $\Psi(\underset{\sim}{\lambda})$.

Finally two other results will be mentioned which give intuition to the choice of $\underset{\sim}{\lambda}$ or $\underset{\sim}{\theta}$

parameters in these penalty functions. Both show that minimizing a penalty function is equivalent to solving a constrained minimization problem which is a perturbation of the original problem. For the equality problem Powell (1969) showed that

Theorem 3 $\underset{\sim}{x}(\underset{\sim}{\lambda})$ is a strong local minimizer of the problem: minimize $F(\underset{\sim}{x})$ subject to $\underset{\sim}{c}(\underset{\sim}{x}) = \underset{\sim}{c}(\underset{\sim}{x}(\underset{\sim}{\lambda}))$.

The proof (see Powell) is immediate, and the result emphasizes the need to choose $\underset{\sim}{\lambda}$ so as to make $\underset{\sim}{c}(\underset{\sim}{x}(\underset{\sim}{\lambda})) = 0$. A similar result is demonstrated for the inequality problem.

Theorem 4 $\underset{\sim}{x}(\underset{\sim}{\lambda})$ is a strong local minimizer of the problem: minimize $F(\underset{\sim}{x})$ subject to $c_i(\underset{\sim}{x}) \geq \min(c_i(\underset{\sim}{\lambda}), \theta_i)$ for all i.

Proof By the optimality of $\underset{\sim}{x}(\underset{\sim}{\lambda})$, for any neighbouring $\underset{\sim}{x}$ it follows that $\Phi(\underset{\sim}{x},\underset{\sim}{\lambda}) > \Phi(\underset{\sim}{x}(\underset{\sim}{\lambda}),\underset{\sim}{\lambda})$ and hence

$$F(\underset{\sim}{x}) + \tfrac{1}{2}\sum_i \sigma_i (c_i(\underset{\sim}{x})-\theta_i)_-^2 > F(\underset{\sim}{x}(\underset{\sim}{\lambda})) + \tfrac{1}{2}\sum_i \sigma_i (c_i(\underset{\sim}{x}(\underset{\sim}{\lambda}))-\theta_i)_-^2$$

If now $c_i(\underset{\sim}{x}) \geq \min(c_i(\underset{\sim}{x}(\underset{\sim}{\lambda})),\theta_i)$ for all i, then (i) if $c_i(\underset{\sim}{x}(\underset{\sim}{\lambda})) \geq \theta_i$ then $(c_i(\underset{\sim}{x})-\theta_i)_- = 0 = (c_i(\underset{\sim}{x}(\underset{\sim}{\lambda}))-\theta_i)_-$, and (ii) if $c_i(\underset{\sim}{x}(\underset{\sim}{\lambda})) < \theta_i$ then $(c_i(\underset{\sim}{x}(\underset{\sim}{\lambda}))-\theta_i)_- \leq (c_i(\underset{\sim}{x})-\theta_i)_- \leq 0$. Hence for both (i) and (ii), $F(\underset{\sim}{x}) > F(\underset{\sim}{x}(\underset{\sim}{\lambda}))$ and the Theorem follows. ∎

This time the need to choose $\underset{\sim}{\lambda}$ so as to make $\min(c_i,\theta_i) = 0$ is emphasized.

3. Strategies for changing $\underset{\sim}{\lambda}$ (or $\underset{\sim}{\theta}$)

In this section iterations for correcting the $\underset{\sim}{\lambda}$ parameters will be investigated of the form

$$\lambda^{(k+1)} = \lambda^{(k)} + \Delta\lambda^{(k)} \tag{3.1}$$

where superscripts denote iteration number. Because $\psi(\lambda)$ is twice differentiable it is possible to state a Newton algorithm for this iteration. By virtue of (2.8) this is the correction

$$\Delta\lambda^{(k)} = -(N^T G^{-1} N)^{-1} c \tag{3.2}$$

where $N = N(x(\lambda^{(k)}))$ etc. For an inequality constraint problem, by virtue of (2.14) and (2.15) the correction (3.2) is appropriate for the multipliers λ_i $i \in M(\lambda^{(k)})$, together with the correction

$$\Delta\lambda_i^{(k)} = -\lambda_i^{(k)} \quad i \notin M(\lambda^{(k)}) \tag{3.3}$$

for the other multipliers. However the correction (3.2) does not necessarily keep $\lambda \geq 0$, and in view of the necessary condition (1.12a) a probably more effective iteration is to choose $\lambda^{(k+1)}$ as the minimizer of the subproblem

$$\text{minimize} \quad Q^{(k)}(\lambda) \quad \text{subject to } \lambda \geq 0 \tag{3.4}$$

where

$$Q^{(k)}(\lambda) = \tfrac{1}{2}(\lambda-\lambda^{(k)})^T \left\{\nabla_\lambda^2 \Psi(\lambda^{(k)})\right\} (\lambda-\lambda^{(k)}) + (\lambda-\lambda^{(k)})^T \nabla_\lambda \Psi(\lambda^{(k)})$$

comprises the first and second terms of the Taylor expansion of $\Psi(\lambda)$ about $\lambda^{(k)}$. In fact the solution of (3.4) for the multipliers λ_i $i \notin M(\lambda^{(k)})$ is also given by (3.3), so it is only necessary to pose (3.4) in terms of the multipliers λ_i $i \in M(\lambda^{(k)})$, and there are usually not more than n elements in this set. The subproblem (3.4) is a quadratic program with only simple lower bounds and can be solved efficiently (Fletcher and Jackson, 1973).

If exact values of $\nabla_{\lambda}\Psi$ and $\nabla^2_{\lambda}\Psi$ are available, and if convergence of (3.2) to the optimum parameters λ^* does occur, then the rate of convergence is known to be second order. This is also true for (3.4) assuming that $A^* \equiv \bar{A}^*$ (strict complementarity) because ultimately the bounds $\lambda_i \geq 0$ $i \in M(\lambda^{(k)})$ are not active. However I would expect a second order rate of convergence to hold even when strict complementarity does not occur.

In view of the fact that λ^* is the unconstrained maximum of $\Psi(\lambda)$ however, there is no necessity to solve the quadratic program (3.4). The Newton iteration based on (3.2) and (3.3) is more simple and avoids the need for a routine for (3.4). It would also be possible to modify the solution to (3.2) to ensure $\lambda \geq 0$ by any convenient ad-hoc rule, and the resulting algorithm would probably not be much inferior to (3.4).

It is well known that a disadvantage of Newton's method is that both first and second derivatives of the objective function must be available. The same is true here because to evaluate $\nabla^2_{\lambda}\Psi$ requires the matrix $G = \nabla^2\Psi(x(\lambda),\lambda)$ which involves second derivatives of the functions $F(x)$ and $c_i(x)$. Nevertheless most good unconstrained minimization routines build up estimates of G (or G^{-1}) either directly as in quasi-Newton methods, or indirectly as in conjugate direction methods. Thus if this estimate is used in

(3.2) or (3.4) a Newton-like algorithm results for which a rapid rate of convergence is likely even though only first derivatives of the functions F and c_i are required.

An even more simple correction formula is suggested for the equality problem by Powell (1969) and by Hestenes (1969). It will be shown simply how this formula can be derived from (3.2) and also what the corresponding formula is for the inequality problem. Consider the matrix $G^*(S) = \nabla^2\psi(\underset{\sim}{x}^*,\underset{\sim}{\lambda}^*,S)$ and let S be increased by adding a large positive diagonal matrix D. Then

$$G^*(S+D) = G^*(S) + N^*DN^{*T} \tag{3.5}$$

follows from the definition of ψ, (1.4). Use of the well known formula for the correction to the inverse matrix (see Householder, 1964, for example) enables an expression for $[G^*(S+D)]^{-1}$ to be obtained. By applying the same formula again it follows that

$$(N^{*T}[G^*(S+D)]^{-1}N^*)^{-1} = (N^{*T}[G^*(S)]^{-1}N^*)^{-1} + D. \tag{3.6}$$

Consider now the iteration (3.1) and let the correction be determined by

$$\underset{\sim}{\Delta}\underset{\sim}{\lambda}^{(k)} = -A^{(k)}\underset{\sim}{c}^{(k)} .$$

The resulting iteration is superlinearly convergent if $A^{(k)} \to (N^{*T}G^{*-1}N^*)^{-1}$. Furthermore linear convergence at any desired rate can be achieved if $\lim_{k\to\infty} A^{(k)}$ is sufficiently close to $(N^{*T}G^{*-1}N^*)^{-1}$, relatively speaking. Now if all the elements of D are large relative to S, then it follows that

$$(N^{*T}[G^*(S+D)]^{-1}N^*)^{-1} = (S+D) + 0(S). \tag{3.7}$$

That is to say, if S ı-3+D is made sufficiently large, then arbitrarily good relative agreement between S and $[\nabla_\lambda^2 \Psi]^{-1}$ is obtained. If $A^{(k)}$ is replaced by S then the correction becomes

$$\Delta\underset{\sim}{\lambda}^{(k)} = - S\underset{\sim}{c}^{(k)} \,, \tag{3.8a}$$

or in terms of the $\underset{\sim}{\theta}$ parameters

$$\Delta\underset{\sim}{\theta}^{(k)} = - \underset{\sim}{c}^{(k)}, \tag{3.8b}$$

and these are the simple formulae suggested by Hestines and Powell respectively. One especial merit is that the formulae do not require the derivatives N and G and therefore are suitable when the minimization routine applied in $\underset{\sim}{x}$-space does not require derivatives. An equally simple correction formula holds for the inequality problem. Equation (3.7) shows that the quadratic function $Q^{(k)}(\underset{\sim}{\lambda})$ in (3.4) tends to a sum of squares as $k \to \infty$ and $S \to \infty$. Using this fact to determine the correction yields the formula

$$\Delta\theta_i^{(k)} = - \min(c_i^{(k)}, \theta_i^{(k)}) \quad i = 1,2,\ldots,m, \tag{3.9a}$$

or in terms of the $\underset{\sim}{\lambda}$ parameters,

$$\Delta\lambda_i^{(k)} = - \min(\sigma_i^{(k)} c_i^{(k)}, \lambda_i^{(k)}), \quad i = 1,2,\ldots,m \tag{3.9b}$$

In section 5 some numerical results are given comparing the Newton-like iteration based on (3.4) against the Powell/Hestenes iteration (3.9).

The Newton correction (3.2) has also been suggested by Buys when second derivatives are explicitly available, but he does not seem to consider using approximations to G. Furthermore he concludes (wrongly in my view - see (2.15)) that $\nabla^2_\lambda \Psi(\underset{\sim}{\lambda})$ is always singular and that this precludes the use of Newton-like algorithms for varying $\underset{\sim}{\lambda}$ in the inequality problem. In my opinion (backed up by the numerical experience of section 5) the local rate of convergence of these algorithms is very satisfactory.

4. Convergence; variation of S

So far attention has only been given to local convergence reuslts for iterative methods which find the optimum multipliers $\underset{\sim}{\lambda}^*$. In this section consideration will be given to ways of obtaining guaranteed convergence by varying the elements of S. In fact under certain conditions it is possible to prove convergence of (3.8) without increasing S at all, providing that Theorem 1 is satisfied. To do this requires an inequality given by Rockafellar for the case $S = \sigma I$, that for all $\underset{\sim}{\lambda}, \underset{\sim}{\lambda}'$

$$\Psi(\underset{\sim}{\lambda}') \geq \Psi(\underset{\sim}{\lambda}) + (\underset{\sim}{\lambda}'-\underset{\sim}{\lambda})^T \underset{\sim}{\nabla}_\lambda \Psi(\underset{\sim}{\lambda}) - \tfrac{1}{2}\sigma^{-1}(\underset{\sim}{\lambda}'-\underset{\sim}{\lambda})^T(\underset{\sim}{\lambda}'-\underset{\sim}{\lambda}).$$

This inequality shows the matrix $-\nabla^2_\lambda \Psi$ is bounded away from zero by $\sigma^{-1}I$. However the derivation of this inequality depends strongly on convexity assumptions about the original problem, and I am not clear to what extent such conclusions are valid in general.

Most global convergence results make the assumption that $\underset{\sim}{x}(\underset{\sim}{\lambda})$ is the <u>global</u> minimizer of the penalty function. It is not known to what extent this is inevitable, and it would be of interest to conduct a study into this problem. The role of S in forcing convergence is illustrated by the result that if

$\sigma_i^{(k)} \to \infty$ and $\underset{\sim}{\lambda}$ is fixed, then $c_i^{(k)} \to 0$, and behaves asympotically like $c_i^{(k)} \sim \text{const}/\sigma_i^{(k)}$. Thus it would be expected that a sensible way of using S is to increase S so as to force the iterates $\underset{\sim}{\lambda}^{(k)}$ into a region about $\underset{\sim}{\lambda}^*$ in which the local convergence results are valid. Once in this region then S stays fixed and $\underset{\sim}{\lambda}$ alone is varied, and $\underset{\sim}{\lambda} \to \underset{\sim}{\lambda}^*$ at the appropriate rate.

A suitable way of achieving this aim is suggested by Powell (1969) in regard to the equality problem (1.1). He measures convergence in terms of

$$K = \|\underset{\sim}{c}(\underset{\sim}{x}(\underset{\sim}{\lambda}),S)\|_\infty . \tag{4.1}$$

If an iteration is deemed to be completed when a better value of K is found, and if the corresponding $\underset{\sim}{\lambda}$,S are denoted by $\underset{\sim}{\lambda}^{(k)}, S^{(k)}$, then Powell's method is the following.

Initially set $\underset{\sim}{\lambda}=\underset{\sim}{\lambda}^{(1)}$, $S=S^{(1)}$, $k=0$, $K^{(0)}=\infty$.

(i) Evaluate $\underset{\sim}{x}(\underset{\sim}{\lambda},S)$ and $\underset{\sim}{c} = \underset{\sim}{c}(\underset{\sim}{x}(\underset{\sim}{\lambda},S))$. (4.2)

(ii) Find $\{i:|c_i|\geq K^{(k)}/4\}$. If $\|\underset{\sim}{c}\|_\infty \geq K^{(k)}$ go to (v).

(iii) Set $k=k+1$, $\underset{\sim}{\lambda}^{(k)}=\underset{\sim}{\lambda}$, $S^{(k)}=S$, $K^{(k)}=\|\underset{\sim}{c}\|_\infty$. Finish if $K^{(k)} \leq \varepsilon$.

(iv) If $K^{(k)}\leq K^{(k-1)}/4$ or $\underset{\sim}{\lambda}^{(k)}=\underset{\sim}{\lambda}^{(k-1)}$,
(set $\underset{\sim}{\lambda}=\underset{\sim}{\lambda}^{(k)}-S^{(k)}c^{(k)}$, go to (i)).

(v) Set $\underset{\sim}{\lambda}=\underset{\sim}{\lambda}^{(k)}$, $\sigma_i=10\sigma_i$ $\forall$ $i\varepsilon\{i\}$, go to (i).

The test $\underset{\sim}{\lambda}^{(k)}=\underset{\sim}{\lambda}^{(k-1)}$ in (iv) is equivalent to testing whether control last flowed into (i) from (v). This iteration is chosen so that ultimately $K^{(k)}$ is reduced by a factor of at least 1/4 at each iteration. Although an early iteration may involve more than one unconstrained minimization of the penalty function, each iteration cannot fail to terminate. Powell proves convergence of his iteration in that the test $K^{(k)} \leq \varepsilon$ must ultimately be satisfied. To do this requires the global result (2.9), and also the assumption that F(x) is bounded below, although this latter assumption can be dispensed with. A similar process to (4.2) can be used to force convergence of the Newton iteration (3.2), merely by an appropriate change of the formula $\underset{\sim}{\lambda}=\underset{\sim}{\lambda}^{(k)}-S^{(k)}\underset{\sim}{c}^{(k)}$ in (4.2(iv)).

An important observation is that the minimizer $\underset{\sim}{x}(\underset{\sim}{\lambda},S)$ of the penalty function $\psi(\underset{\sim}{x},\underset{\sim}{\lambda},S)$ could have been obtained from any one of an infinity of values of $\underset{\sim}{\lambda}$,S. If $\underset{\sim}{\lambda}'$,S' is one of these values, then it is only necessary for $\underset{\sim}{\lambda}'$,S' to be related to $\underset{\sim}{\lambda}$,S by

$$\underset{\sim}{\lambda}' - S'\underset{\sim}{c} = \underset{\sim}{\lambda} - S\underset{\sim}{c} \tag{4.3}$$

where $\underset{\sim}{c} = \underset{\sim}{c}(\underset{\sim}{x}(\underset{\sim}{\lambda},S)) = \underset{\sim}{c}(\underset{\sim}{x}(\underset{\sim}{\lambda}',S'))$. This is by virtue of the condition

$$\underset{\sim}{\nabla}\psi(\underset{\sim}{x}(\underset{\sim}{\lambda},S),\underset{\sim}{\lambda},S) = \underset{\sim}{\nabla}F(\underset{\sim}{x}(\underset{\sim}{\lambda},S)) - N(\underset{\sim}{\lambda}-S\underset{\sim}{c}(\underset{\sim}{x}(\underset{\sim}{\lambda},S))) = 0$$

which remains satisfied when the change (4.3) is made. It is possible therefore to change $\underset{\sim}{\lambda}$,S subject to (4.3) after a minimization is completed (subject to S being sufficiently large so that $\nabla^2\psi$ is still positive definite) and it is interesting to observe what happens if the subsequent iteration is carried out in terms of the new values $\underset{\sim}{\lambda}'$,S'. In fact because $\underset{\sim}{\lambda}-S\underset{\sim}{c}$ remains constant, and becuase the new iterate

$\lambda^{(k+1)}$ is just λ-Sc, it is easy to see that $\lambda^{(k+1)}$ remains unaffected when the Powell/Hestenes correction is used on the values λ',S'. The same is also true when $\lambda^{(k+1)}$ is determined by Newton's method (3.2) although the analysis is less obvious. This must be so however because the Newton prediction is exact for some problems and therefore shold be independent of how it is predicted. The only effect of making the change (4.3) then is that the value of $S^{(k+1)}$ which is used with $\lambda^{(k+1)}$ is changed. Therefore it is quite consistent when devising an algorithm, to make the change to λ by virtue of (3.2) or (3.8) based on $\lambda^{(k)}$,$S^{(k)}$, and then to choose $S^{(k+1)}$ arbitrarily larger than $S^{(k)}$. An algorithm in which both are changed simultaneously is more efficient because it avoids an evaluation of $x(\lambda,S)$ which would be needed if the changes were made separately.

An example of this is in the Powell algorithm (4.2) where if $K^{(k-1)} > K^{(k)} > K^{(k-1)}/4$ then λ is kept constant whilst S is increased. In fact it is possible to correct λ and also increase S at the same time, and this modification has been found slightly more efficient. The modified algorithm is

Initially set $\lambda=\lambda^{(1)}$, $S=S^{(1)}$, k=0, $K^{(0)}=\infty$.

(i) Evaluate $x(\lambda,S)$ and $c=c(x(\lambda,S))$. (4.4)

(ii) Find $\{i:|c_i|\geq K^{(k)}/4\}$. If $\|c\|_\infty \geq K^{(k)}$, (set $\lambda=\lambda^{(k)}$, go to (v)).

(iii) Set k=k+1, $\lambda^{(k)}=\lambda$, $S^{(k)}=S$, $K^{(k)}=\|c\|_\infty$, Finish if $K^{(k)} \leq \varepsilon$.

(iv) Set $\underset{\sim}{\lambda}=\underset{\sim}{\lambda}^{(k)}-S^{(k)}\underset{\sim}{c}^{(k)}$, If $K^{(k)} \leq K^{(k-1)}/4$ go to (i).

(v) Set $\sigma_i=10\sigma_i \ \forall \ i \ \varepsilon \ \{i\}$, go to (i).

Some results are given in section 5 for this algorithm, and also for one in which a Newton-like correction is used in (4.4 (iv)).

The algorithm (4.4) is valid only for the equality problem but a small modification enables it to be used for the inequality problem. To do this the definition of K (4.1) is changed so that

$$K = \max_i |\min(c_i(\underset{\sim}{\lambda},S),\theta_i)|$$

and a similar change is made where $||c||_\infty$ occurs in (4.4). Also the correction formula to be used must be one appropriate to the inequality problem (that is (3.4) or (3.9)). Of course it is possible to solve problems with mixed equality and inequality constraints and the modifications to do this are similar.

When using either the Powell/Hestenes formula (3.8) or the Newton-like formula based on (3.2) for changing $\underset{\sim}{\lambda}$ inside an algorithm like (4.4), it will be noticed that with the Newton-like method S is only increased so as to force global convergence, whereas with the Powell/Hestenes formula it may be increased further to force the sufficiently rapid rate of linear convergence to $\underset{\sim}{\lambda}^*$. Thus it would be expected that the values of σ_i used by the Newton-like method would be smaller than those used by the Powell/Hestenes method. This has been borne out in practice. Intuitively this would appear to be good in that too large a value of S might make the minimum $\underset{\sim}{x}(\lambda, S)$ difficult to determine due to ill-conditioning of the penalty function. However no such evidence has been forthcoming on the variety of problems considered, and indeed some advantages of having S larger have showed up. One of these of course is that the larger S is, so the more rapidly is the region reached in which the local convergence results are valid. Furthermore the basic strategy of (4.2) and (4.4) is dependent on the asymptotic result $c_i \sim \text{const}/\sigma_i$ and this is only valid for large σ_i. Although clearly the optimum values of S must be a balance between these effects, practical experience indicates that the values of S chosen when using the Powell/Hestenes correction formula in (4.4) have not caused any loss of accuracy in $\underset{\sim}{x}(\underset{\sim}{\lambda}, S)$.

Two observations from the numerical evidence of section 5 suggest a further algorithm for changing S. It is noticeable that the rapid rate of convergence of the Newton-like iteration is significant in reducing the overall number of iterations when there is no difficulty in getting convergence to occur. On the other hand the values of S required to get the appropriate rate of linear convergence with the Powell/Hestenes formula, are always sufficient to force global convergence. Therefore a Newton-like algorithm has been investigated in which S is chosen so that the predicted solution of the Powell/Hestenes iteration would reduce $K^{(k)}$ to $K^{(k)}/4$. Let $\underset{\sim}{\Delta}^{PH}$ and $\underset{\sim}{\Delta}^{N}$ be the corrections predicted by the Powell/Hestenes and Newton formulae respectively, and $\underset{\sim}{\lambda}^{PH}$ and $\underset{\sim}{\lambda}^{N}$ the corresponding prediction of the multipliers. Assume $\psi(\underset{\sim}{\lambda})$ to be quadratic, so that $\underset{\sim}{\lambda}^N = \underset{\sim}{\lambda}^*$, whence because $\underset{\sim}{c}(\underset{\sim}{\lambda}^*) = 0$,

$$(N^T G^{-1} N)^{-1} \underset{\sim}{c}(\underset{\sim}{\lambda}^{PH}) = \underset{\sim}{\lambda}^{PH} - \underset{\sim}{\lambda}^* = \underset{\sim}{\Delta}^{PH} - \underset{\sim}{\Delta}^N. \qquad (4.5)$$

The right hand side of (4.5) is independent of S, so the effect of increasing S is to increase $(N^T G^{-1} N)^{-1}$ and hence to decrease $\underset{\sim}{c}(\underset{\sim}{\lambda}^{PH}, S)$. For large, S, $(N^T G^{-1} N)^{-1}$ can be estimated by S (see 3.7 for instance). Now S is to be found so that

$$|c_i(\underset{\sim}{\lambda}^{PH}, S)| \le |c_i(\underset{\sim}{\lambda}^{(k)}, S^{(k)})|/4 = |\Delta_i^{PH}|/(4\sigma_i^{(k)})$$

by (3.8). Hence from (4.5) it follows that

$$\sigma_i \ge 4\sigma_i^{(k)} \left| \frac{\Delta_i^{PH} - \Delta_i^N}{\Delta_i^{PH}} \right| . \qquad (4.6)$$

An algorithm based on using this formula to choose σ_i at each iteration has also been tried. In this algorithm σ_i is only increased by the factor 10 if the $\underset{\sim}{\lambda}$ correction formula fails to improve K. The algorithm is

Initially set $\underset{\sim}{\lambda} = \underset{\sim}{\lambda}^{(1)}$, $S = S^{(1)}$, $k=0$, $K^{(0)} = \infty$.

(i) Evaluate $\underset{\sim}{x}(\underset{\sim}{\lambda}, S)$ and $\underset{\sim}{c} = \underset{\sim}{c}(\underset{\sim}{x}(\underset{\sim}{\lambda}, S))$. (4.7)

(ii) If $\|\underset{\sim}{c}\|_\infty \ge K^{(k)}$, (set $\sigma_i = 10\sigma_i$ $\forall$ i: $|c_i| \ge K^{(k)}$, go to (i))

(iii) Set $k=k+1$, $\underset{\sim}{\lambda}^{(k)} = \underset{\sim}{\lambda}$, $S^{(k)} = S$, $K^{(k)} = \|\underset{\sim}{c}\|_\infty$, Finish if $K^{(k)} \le \varepsilon$.

(iv) Set $\underset{\sim}{\lambda}$ by (3.2), increase each σ_i if necessary so as to satisfy (4.6), go to (i).

Numerical results for this algorithm are also described in section 5.

These ideas by no means exhaust the possibilities for a strategy for changing S, and in particular no algorithms have been tried in which the value of $\Psi(\underset{\sim}{\lambda},S)$ is used. Yet this information is readily available, so further research in this direction might be fruitful.

5. Practical experience and discussion

In this section numerical experience gained with the algorithm (4.4) will be described, using both the Powell/Hestenes formulae (3.8, 3.9) and the Newton-like formula based on (3.2) and (3.4) to change the $\underset{\sim}{\lambda}$ parameters. Experience with the algorithm (4.7) is also described. A general program has been written and modified for each of these algorithms, but various features are common to all. The program works with scaled constraint values, that is the user supplies a vector $\underset{\sim}{\bar{c}} > 0$ whose magnitude is typical of that of the constraint functions $\underset{\sim}{c}(\underset{\sim}{x})$. Theprogram then works with constraint functions $\underset{\sim}{c}'(\underset{\sim}{x})$, where $c_i' = c_i/\bar{c}_i$. The initial $\underset{\sim}{\lambda}$ and S are set automatically by the program unless the user chooses otherwise. For instance the user might want to try the choice $\underset{\sim}{\lambda} = S\underset{\sim}{c} + N^+\underset{\sim}{\nabla}F$ which minimizes $\|\underset{\sim}{\nabla}\phi\|_2$. The automatic choice of $\underset{\sim}{\lambda}$ is $\underset{\sim}{\lambda}=0$, and the choice for S is based on the following criterion. A rough estimate of the likely change ΔF in F on going to the solution is made by the user. ΔF is used to scale the other terms which occur in the penalty function, so that σ_i is set to make $\frac{1}{2}\sigma_i c_i'^2 = |\Delta F|$. A quasi-Newton method VA09A from the Harwell subroutine library is used to minimize $\Phi(\underset{\sim}{x},\underset{\sim}{\theta},S)$ with respect to $\underset{\sim}{x}$, and the initial estimate of $\nabla^2\Phi$ can either be set automatically to I, or otherwise by the user. However the suggestion by Buys (1972) that the estimate be reset to the unit matrix whenever the active set is changed is not used, and in my opinion would be rather inefficient. In fact the approximation to $\nabla^2\Phi$ is carried forward from one minimization to the next, and whenever S is changed the estimate of $\nabla^2\Phi$ is changed by virtue of (3.5). This involves a rank one correction to the estimate for every σ_i which is increased. The routine (and also VA09A), uses LDL^T factorizations to represent $\nabla^2\Phi$, and this contributes to the accuracy of the process. Double length computation on an IBM 370/165 computer is used in the tabulated results and the convergence criterion is that $K \leq 10^{-6}$.

A wide selection of test problems has been used. These are the parcel problem (PP) of Rosenbrock (1960), the problem (RS) due to Rosen and Suzuki (1965), the problem (P) due to Powell (1969), and four test problems (TP1,2,3 and 7) used in the comparisons carried out by Colville (1968). The features of these problems are set out in Table 1, where m_i, m_e, and m_a indicate the numbers of inequality, equality and active (at $\underset{\sim}{x}^*$) constraints respectively. P(A) and P(B) etc. indicates that the same problem has been repeated with different initial S values. The criterion used for comparison is the number of times that F, $\underset{\sim}{c}$, $\underset{\sim}{\nabla}F$ and $[\underset{\sim}{\nabla}c_i]$ together are evaluated for given $\underset{\sim}{x}$. In fact however $\underset{\sim}{\nabla}c_i$ is not evaluated for any inequality constraint for which $c_i \geq \theta_i$, and $\underset{\sim}{\nabla}c_i$ for any linear constraint is set on entry to the program as it is a constant vector. Not only has the total number of evaluations on each problem been tabluated but also the number required on iterations after the first. The first minimization is the same in each case and is often the most expensive, and this can obscure the comparison.

The detailed performance of the three different algorithms tested is given in tables 2, 3 and 4. The most striking feature is that the number of outer iterations taken by methods based on the Newton-like formula is far fewer than is taken when using the Powell/Hestenes formula. This substantiates empirically the second-order convergence of these methods, even though the second derivatives $\nabla^2_\lambda \Psi(\underset{\sim}{\lambda})$ are not caluculated exactly. Another pointer to this fact is in the values of K on the later minimizations. For the Powell/Hestenes formula the values go down in a way which appears to be linear and the final K values are all in the range $(10^{-6}, 10^{-7})$. For the Newton-like formula the ratio of successive K values increases for increasing k, suggesting superlinear convergence, and the final K values are often much smaller than 10^{-6}. However the difference in number of evaluations is not as severe as this discrepancy in minimizations might suggest, because the successive minima take fewer evaluations to compute. This is presumably because each starting approximation is closer due to a smaller change being made to the $\underset{\sim}{\lambda}$ parameters. Another feature of interest is that the correct active set for an inequality problem is usually established quickly by the $\underset{\sim}{\lambda}$ iteration. Incidentally it is instructive that it is not worth trying to extrapolate these methods by estimating a starting value of $\underset{\sim}{x}$ for $\Phi(\underset{\sim}{x},\underset{\sim}{\lambda}^{(k+1)})$ from information taken at the solution $\underset{\sim}{x}(\underset{\sim}{\lambda}^{(k)})$. It is merely necessary to choose this starting value as $\underset{\sim}{x}(\underset{\sim}{\lambda}^{(k)})$ because the first step of the quasi-Newton method applied to $\Phi(\underset{\sim}{x},\underset{\sim}{\lambda}^{(k+1)})$ will move $\underset{\sim}{x}^{(1)}$ in the direction of the extrapolated minimum, assuming that an updated estimate of $\nabla^2\Phi(\underset{\sim}{x},\underset{\sim}{\lambda}^{(k+1)})$ has been used.

In examing the problems individually, it is noticeable that when solving TP1 and TP7 the effect of estimating second derivatives of $\nabla^2\Phi$ for the first minimization leads to a particularly good number of evaluations for that minimization. The problem TP3(A) has a poor estimate of the likely change ΔF in $F(x)$ and so the σ_i are estimated very much on the small side. This causes a slow rate of convergence until larger values are obtained.

TABLE 1

Resumé of problems and performance

Problem	Type of Problem				Performance					
	n	m_i	m_e	m_a	Powell		Newton		Mod. Newton	
PP	3	7	-	1	37*	22**	30	15	30	15
RS	4	3	-	1	57	37	36	16	35	15
P(A)	5	-	3	3	45	27	32	14	32	14
P(B)	5	-	3	3	52	36	40	24	37	21
TP1	5	15	-	4	51	36	40	25	39	24
TP2	15	20	-	11	149	21	181	53	162	34
TP3(A)	5	16	-	5	95	70	113	88	101	76
TP3(B)	5	16	-	5	64	33	94	63	64	33
TP7	16	32	8	13	89	73	65	49	53	37

* total number of evaluations

** number of evaluations excepting the first minimiation

TABLE 2

Powell/Hestenes correction (3.9) in algorithm (4.4)

correct active set established (indicated by |)

Minimi- -zation	1	2	3	4	5	6	7	8	9	10
PP	15 .15	| 8 | .014	4 $.1_{10}{}^{-2}$	4 $.1_{10}{}^{-3}$	3 $.1_{10}{}^{-4}$	3 ← no. of evaluations $.8_{10}{}^{-6}$ ← K				
RS	20 .17	| 7 | .033	5 .017	6 $.3_{10}{}^{-2}$	5 $.5_{10}{}^{-3}$	4 $.8_{10}{}^{-4}$	4 $.1_{10}{}^{-4}$	3 $.2_{10}{}^{-5}$	3 $.3_{10}{}^{-6}$	
P (A)	| 18 | .067	7 .011	4 $.2_{10}{}^{-2}$	4 $.3_{10}{}^{-3}$	4 $.5_{10}{}^{-4}$	3 $.8_{10}{}^{-5}$	3 $.1_{10}{}^{-5}$	2 $.2_{10}{}^{-6}$		
P (B)	| 16 | .84	7 .60	8 .20	7 .014	5 $.2_{10}{}^{-2}$	4 $.1_{10}{}^{-3}$	3 $.7_{10}{}^{-5}$	2 $.4_{10}{}^{-6}$		
TP1	15 .45	6 .10	16 .032	| 4 | $.2_{10}{}^{-2}$	4 $.7_{10}{}^{-4}$	3 $.3_{10}{}^{-5}$	3 $.1_{10}{}^{-6}$			
TP2	128 .28	15 $.5_{10}{}^{-4}$	| 6 | $.5_{10}{}^{-6}$							
TP3 (A)	25 .95	5 .48	13 .34	19 .20	7 .22	8 .039	| 6 | $.4_{10}{}^{-2}$	5 $.1_{10}{}^{-3}$	4 $.6_{10}{}^{-5}$	3 $.3_{10}{}^{-6}$
TP3 (B)	31 .27	11 .11	| 7 | .014	5 $.6_{10}{}^{-3}$	4 $.4_{10}{}^{-4}$	3 $.3_{10}{}^{-5}$	3 $.2_{10}{}^{-6}$			
TP7	16 .96	10 .75	11 .30	9 .15	12 .067	12 .010	7 $.4_{10}{}^{-2}$	5 $.9_{10}{}^{-4}$	4 $.7_{10}{}^{-5}$	3 $.6_{10}{}^{-6}$

TABLE 3

Newton-like correction based on (3.4) in algorithm (4.4)

Minimization	1	2	3	4	5	6	7	8	9	10
PP	15 .15	8 $.4_{10}{-2}$	4 $.7_{10}{-5}$	3 $.5_{10}{-9}$						
RS	20 .17	7 $.4_{10}{-2}$	5 $.5_{10}{-4}$	4 $.8_{10}{-6}$						
P(A)	18 .067	7 $.2_{10}{-2}$	4 $.1_{10}{-4}$	3 $.1_{10}{-7}$						
P(B)	16 .84	9 .26	7 .011	5 $.7_{10}{-4}$	3 $.6_{10}{-6}$					
TP1	15 .45	18 .024	4 $.1_{10}{-3}$	3 $.1_{10}{-6}$						
TP2	128 .28	16 .050	19 $.2_{10}{-2}$	9 $.2_{10}{-3}$	5 $.5_{10}{-5}$	4 $.9_{10}{-7}$				
TP3(A)	25 .95	35 2.15	24 .32	11 .12	10 $.3_{10}{-3}$	5 $.5_{10}{-5}$	3 $.1_{10}{-7}$			
TP3(B)	31 .27	38 1.09	13 .040	5 $.5_{10}{-3}$	4 $.2_{10}{-5}$	3 $.1_{10}{-10}$				
TP7	16 .96	15 .30	12 .051	8 $.2_{10}{-2}$	6 $.2_{10}{-3}$	5 $.6_{10}{-5}$	3 $.1_{10}{-6}$			

TABLE 4

Newton-like correction based on (3.4) in algorithm (4.7)

Minimization	1	2	3	4	5	6	7	8	9	10
PP	15 .15	8 $.4_{10^{-2}}$	4 $.7_{10^{-5}}$	3 $.5_{10^{-9}}$						
RS	20 .17	7 $.4_{10^{-2}}$	5 $.8_{10^{-5}}$	3 $.4_{10^{-7}}$						
P(A)	18 .067	7 $.2_{10^{-2}}$	4 $.2_{10^{-4}}$	3 $.2_{10^{-6}}$						
P(B)	16 .84	8 .10	6 $.9_{10^{-3}}$	4 $.2_{10^{-4}}$	3 $.6_{10^{-10}}$					
TP1	15 .45	17 $.7_{10^{-2}}$	4 $.5_{10^{-4}}$	3 $.3_{10^{-6}}$						
TP2	128 .28	16 .050	8 $.3_{10^{-2}}$	6 $.3_{10^{-4}}$	4 $.9_{10^{-6}}$					
TP3(A)	25 .95	29 .23	27 .049	12 $.6_{10^{-3}}$	5 $.3_{10^{-4}}$	3 $.3_{10^{-7}}$				
TP3(B)	31 .27	8 .062	17 $.2_{10^{-2}}$	5 $.2_{10^{-4}}$	3 $.2_{10^{-7}}$					
TP7	16 .96	14 .15	10 .083	6 $.6_{10^{-3}}$	4 $.7_{10^{-5}}$	3 $.4_{10^{-7}}$				

The effect is particularly noticeable with the Newton-like iteration in (4.4). Increasing the initial σ_i by 10 (TP3(B)) improved matters considerably. However for problem TP7 the initial S is adequate to ensure convergence and here the advantage of the Newton-like iteration is most apparent. The problem TP2 is anomalous in that the Powell/Hestenes formula gives multipliers correct to three figures after one minimization. In view of the results on other problems it seems likely that some special effect may be at work, perhaps on account of TP2 being the dual of the linearly constrained problem TP1.

Overall the best method is that of algorithm (4.7). It is interesting that because this method usually avoids increasing the σ_i by arbitrary factors of 10, it tends to scale the σ_i amongst themselves rather better than the other methods. This method never fails to reduce K and the worst iteration is the one in which K is reduced from .15 to .083. A subroutine VF01A/AD which implements this method is available in the Harwell subroutine library and those interested should contact the subroutine librarian.

When comparing this method against other penalty or barrier functions it is found that the new function has a number of good properties which are not found together in any other penalty function. One of these is good conditioning of Φ due to the fact that no singularities are introduced in the penalty term, and that it is not necessary to make the parameters $\sigma_i \to \infty$ in order to force local convergence. Once S has been made sufficiently large, convergence of the $\underset{\sim}{\lambda}$ iteration occurs at a rapid rate, and numerical experience suggest that high accuracy can be obtained in very few minimizations. Furthermore because the hessian matrix $\nabla^2\Phi$ can be carried forward from one iteration to the next, and updated when necessary, the computational effort required for the successive minimizations goes down rapidly. Most important of all for a penalty or barrier function is that it is very easy to program the method by incorporating an established quasi-Newton minimization routine into the program. With a barrier function, difficult decisions have to be taken about how to define the barrier function in the infeasible region, and it is not easy to avoid having to modify the minimization routine. Furthermore the linear search in the quasi-Newton subroutine is usually based on a cubic interpolation and is unsuitable for functions with singularities. In the Powell/Hestenes/Rockafellar penalty function however the function is defined for all $\underset{\sim}{x}$ and the cubic linear search is also adequate. Finally there is no need to supply an initial feasible point to start off the whole process.

Osborne and Ryan (1972) give a method which also adapts the Powell/Hestenes penalty function to solve inequality constraint problems. Theirs is a hybrid method in which a barrier function is used to get an estimate of the likely active set so that the Powell/Hestenes function can be used, treating this set as equalities. They compare their method against more conventional types of barrier function on a number of problems including the problems TP1, 2, 3 used here. These results enable a general comparison amongst the penalty and barrier functions to be made. Osborne and Ryan work to an accuracy of 10^{-8}, so in comparing their results with those in this paper a small adjustment should be made. Assuming that 1 extra evaluation per minimization and also one extra minimization would be required on to the totals of table 4 to achieve the slightly higher accuracy, the comparison is shown in table 5. These results show a measurable bias in favour of the new penalty function.

TABLE 5
Comparison of penalty and barrier functions

Problem	Newton-like method (4.7)	Osborne and Ryan	Barrier function	Extrapolated B.F.
TP1	47	167	225	177
TP2	172	229	440	245
TP3(B)	73	107	173	123

So far the emphasis has been on the advantages of the new penalty function and it is advisable to consider what the disadvantages are if any. One possible disadvantage is that the presence of discontinuities in the second derivative of the penalty function might cause slow convergence of the quasi-Newton subroutine. An experiment has been conducted to test this hypothesis. The first minimization of $\Phi(\underset{\sim}{x},\underset{\sim}{\theta},S)$ for TP2 was repeated, designating the constraints known to be active at the minimum as equalities. This removes the discontinuities for these constraints and should lead to faster convergence under the hypothesis. In fact three more evaluations were required. The run in which the discontinuities were present was also checked to see whether the discontinuities were active in the sense that points either side of them were being taken, and this was certainly true. Also the results from the other test problems are by no means unduly large for the size of problem involved. Therefore I have no evidence to support the hypothesis that these discontinuities at all retard convergence.

Another possible disadvantage of the penalty function is that if S is not chosen large enough, the local minimum of $\Phi(\underset{\sim}{x},\underset{\sim}{\theta},S)$ at $\underset{\sim}{x}^*$ may not exist, and even if it does, Φ may be unbounded below elsewhere (for example, minimize $-x^3$ subject to $x \leq 1$, for which $\Phi \sim -x^3 + \frac{1}{2}\sigma x^2$ for large x). However there are various ways to get round this, for instance by increasing S, or by replacing $F(\underset{\sim}{x})$ by $\exp(F(\underset{\sim}{x}))$. A related disadvantage is that the method does not appear to handle problems with inequality constraints which cut out regions in which $F(\underset{\sim}{x})$ is not defined. However I have yet to see a problem for which it is not possible to define a smooth continuation of $F(\underset{\sim}{x})$ into the infeasible region. Such a device enables the method to be applied to this type of problem. One related advantage of the method is that it is not necessary for the initial $\underset{\sim}{x}$ approximation to be feasible. However it is interesting to consider what happens when the problem has no feasible point. In this case $K^{(k)}$ is monotonic decreasing but not to zero. Therefore in the equality problem there will be subsequences of {k} for which $\underset{\sim}{c}(\underset{\sim}{\lambda}^{(k)}, S^{(k)}) \to \underset{\sim}{c}'$ where $\underset{\sim}{c}'$ is a fixed vector, and by theorem 3 $\underset{\sim}{x}(\underset{\sim}{\lambda}^{(k)}, S^{(k)})$ will tend to the point $\underset{\sim}{x}'$ which minimizes $F(\underset{\sim}{x})$ subject to $\underset{\sim}{c}(\underset{\sim}{x}) = \underset{\sim}{c}'$. If in fact $\underset{\sim}{c}^{(k)} \to \underset{\sim}{c}'$ for the sequence {k} itself, then it follows in addition that $\underset{\sim}{x}(\underset{\sim}{\lambda}^{(k)}, S^{(k)}) \to \underset{\sim}{x}'$.

A further disadvantage of the penalty function, common to many approaches to the nonlinear programming problem, is that the theory given here calls for an exact local minimum of $\Phi(\underset{\sim}{x}, \underset{\sim}{\theta}, S)$ with respect to $\underset{\sim}{x}$ to be found. Of course in general this cannot be done in a finite number of operations. Now it would be expected that if the theory were modified to remove this necessity, then to get an equivalent rate of convergence would require increasing accuracy in $\underset{\sim}{x}(\underset{\sim}{\lambda}, S)$ to be obtained as $\underset{\sim}{\lambda} \to \underset{\sim}{\lambda}^*$. In practice however most of the effort in unconstrained minimization goes into locating the neighbourhood of the minimum in which fast convergence occurs, and very little into achieving that fast convergence. Thus it costs little in practice to assume that high accuracy in all local minima is obtained, and in VF01A/AD the same accuracy required in the final solution is asked for in all local minima.

In view of this discussion, my opinion is that the disadvantages of the Powell/Hestenes/Rockafellar penalty function are negligible, and that the advantages are strong, especially the lack of numerical difficulties and the ease of using the unconstrained minimization routine. Nonetheless to solve a nonlinear problem by transforming it to a sequence of nonlinear problems should not be optimum, and I think that ultimately algorithms which vary both $\underset{\sim}{x}$ and $\underset{\sim}{\lambda}$ together will prove superior. This is already true as regards local convergence as some results about Solver-like methods which will appear in the thesis of Jackson (1974) show. However a good way of forcing global convergence for such methods is not yet clear.

Finally some possible extensions of the idea are discussed, and in particular the situation vis-a-vis problems in which some of the inequality constraints are linear. For small to medium problems there is not much to be gained by trying to take any special account of this feature, other than to use the fact that $\underset{\sim}{\nabla}c$ is constant and can be set in advance. However for large problems with many active linear and possible sparse constraints, it is worthwhile looking to construct a penalty function from the nonlinear constraints only, to be minimized by a method which maintains feasibility with regard to the linear constraints. In this case it will be important for maximum efficiency to choose a method such as that of Buckley (1973) which only keeps a second derivative approximation in an n-p dimensional space where p is the number of active linear constraints.

Another intersting modification of the method is for problems with two-sided constraints like $a_i \le c_i(\underset{\sim}{x}) \le b_i$. At the moment these must be written as two separate contraints. However this is wasteful in storage space and also there are some implicit restrictions on the $\underset{\sim}{\theta}$ parameters which are not taken account of when separating the two inequalities. A method which keeps the constraints together would be preferable, and would just be an extension of the current method in respect of the way equality constraints are treated, for these are two-sided constraints for which $a_i = b_i$.

References

Arrow, K.J., Gould, F.J. and Howe, S.M., (1973), "A general saddle point result for constrained optimization," Mathematical Programming, Vol. 5, pp. 225-234.

Buckley, A., (1973), "An implementation of Goldfarb's minimization algorithm", AERE report, TP 544.

Buys, J.D., (1972), "Dual algorithms for constrained optimization problems," Ph.D. Thesis, University of Leiden.

Colville, A. R., (1968), "A comparative study on non-linear programming codes," IBM New York Scientific Center Report 320-2949.

Fiacco, A.V. and McCormick, G.P., (1968), "Nonlinear Programming: Sequential Unconstrained Minimization Techniques", Wiley, New York.

Fletcher, R. and Jackson, M.P., (1973), "Minimization of a quadratic function of many variables subject only to lower and upper bounds", AERE report TP 528, and in Jour. Inst. Math. Applics., Vol. 15, pp. 159-174, (1974).

Hestenes, M.R., (1966), "Calculus of variations and optimal control problems," Wiley, New York.

Hestenes, M.R., (1969), "Multiplier and Gradient methods," Jour. Opt. Theory. Appl., Vol. 4, pp. 303-320.

Householder, A.S. (1964) "The theory of matrices in numerical analysis", Blaisdell, New York.

Jackson, M.P., (1974), Ph.D. Thesis, University of Oxford, in preparation.

Mangasarian, O.L., (1973), "Unconstrained Lagrangians in nonlinear programming," University of Wisconsin, Comp. Sci. Tech. Report 174.

Osborne, M.R. and Ryan, D.M., (1972), "A hybrid algorithm for nonlinear programming," Chapter 28 in "Numerical methods for nonlinear optimization," ed. F.A. Lootsma, Academic Press, London.

Powell, M.J.D., (1969), "A method for nonlinear constraints in minimization problems," Chapter 19 in "Optimization", ed. R. Fletcher, Academic Press, London.

Rockafellar, R.T., (1973a), "The multiplier method of Hestenes and Powell applied to convex programming," Jour. Opt. Theory. Appl., Vol. 12, pp. 553-562.

Rockafellar, R.T., (1973b), "A dual approach to solving nonlinear problems by unconstrained optimization," Mathematical Programming, Vol. 5, pp. 354-373.

Rockafellar, R.T., (1974), "Augmented Lagrange multiplier functions and duality in nonconvex programming," SIAM J. Control, Vol. 12, pp. 268-285.

Rosen, J. B. and Suzuki, S., (1965), "Construction of nonlinear programming test problems", Comm. ACM, Vol. 8, p. 113.

Rosenbrock, H.H., (1960), "An automatic method for finding the greatest or least value of a function," Computer J., Vol. 3, pp. 175-184.

Note in Proof. There are some small differences between this paper and that published in the Journal of the Institute of Mathematics and its Applications (Vol. 15, 1975). The latter version states in more detail the consequences of the implicit function theorem, especially for the inequality penalty function. Some of the differentiability properties are also explained more fully. An additional table is given, showing how the parameters λ_i, σ_i and c_i change for the problem TP3.

ON PENALTY AND MULTIPLIER METHODS FOR CONSTRAINED MINIMIZATION*

by

Dimitri P. Bertsekas

ABSTRACT

The purpose of this paper is to present an analysis and a comparison of penalty and multiplier methods for constrained minimization. Global convergence and rate of convergence results are given which show that multiplier methods alleviate to a substantial extent the traditional disadvantages of penalty methods (ill-conditioning, slow convergence), while retaining all of their attractive features. At the same time a global duality framework is constructed in the absence of convexity. Within this framework multiplier methods may be viewed as gradient methods for maximizing a certain dual functional. This interpretation leads to sharper rate of convergence results and motivates efficient modifications of the multiplier iteration.

1. Introduction

The idea of approximating a constrained optimization problem by an unconstrained problem

*This work was supported by the Joint Services Electronics Program (U.S. Army, U.S. Navy, U.S. Air Force) under Contract DAAB-07-72-C-0259.

through the use of a penalty function is a fairly old one. It has been systematically employed in numerical optimization for approximately fifteen years and has been popularized mainly through the work of Fiacco and McCormick [7]. It is accurate to say that penalty function methods have been widely accepted in practice despite criticism directed at their slow convergence properties and the numerical instabilities associated with them. The main reasons for their success are their generality and simplicity relative to other methods for constrained minimization coupled with the fact that they make full use of the very efficient unconstrained minimization procedures which have been developed in recent years. Around 1968 Hestenes [9] and Powell [18](see also [8]) independently proposed a method for constrained minimization, called the method of multipliers, which shared several of the characteristic features of penalty methods. Although the original papers offered limited interpretation or analysis, considerable subsequent research and computation has indicated that the method of multipliers offers significant advantages over penalty methods.

The purpose of this paper is to present analysis related to two alternative ways for viewing the method of multipliers. The first viewpoint is to consider the method of multipliers and the quadratic penalty function method within a common generalized penalty function framework. In the next section we present some convergene estimates for the generalized class of penalty methods for the case of both exact and inexact unconstrained minimization. These estimates firmly establish the advantage of the method of multipliers over the standard quadratic penalty method in terms of speed of convergence. In addition they provide global convergence results for the method of multipliers which represent a substantial improvement over existing local results. The second viewpoint presented is to consider the method of multipliers as a gradient method for maximizing a certain dual functional. This primal-dual viewpoint rests on a global duality framework within which duality gaps due to lack of convexity are eliminated through the convexification effect introduced by the penalty function. The duality theory is presented in Section 3 and the construction is based on the analysis of Section 2. The dual functional constructed has the remarkable properties that it is concave, everywhere finite, everywhere continuously differentiable and twice continuously differentiable over an arbitrary bounded open set S . Furthermore its value and derivatives within S can be obtained by unconstrained minimization of a certain augmented Lagrangian.

For simplicity of presentation we consider equality constraints only. Inequality constraints can be treated without loss of computational efficiency by converting them to equality constraints using slack variables. This device, due to Rockafellar [20], is briefly discussed in Section 4 together with some aspects of the method of multipliers in the presence of convexity.

The presentation is based on an analysis contained in a series of papers and reports by the author [2] - [5]. However several other authors, most notably Buys [6] and Rockafellar [19] - [24], have obtained related results and connections with past and concurrent work are pointed out in this paper and in [2], [3]. No attempt is made to provide a comprehensive survey of the literature and the reader will find many additional references in the papers listed here as well as in the survey papers by Rockafellar [23], [24].

II. A Class of Generalized Penalty Function Methods

Consider the nonlinear programming problem

$$\begin{aligned} &\text{minimize } f(x) \\ &\text{subject to } h(x) = 0 \end{aligned} \tag{1}$$

where $f:R^n \to R$ and $h:R^n \to R^m$. Let $\bar{x}$ be an optimal solution of (1). We assume that f, h are twice continuously differentiable in an open neighborhood of $\bar{x}$ and further adopt the following assumptions:

A.1: The point $\bar{x}$ together with a unique Lagrange multiplier vector $\bar{y} \in R^m$ satisfies the standard second order sufficiency conditions for an isolated local minimum ([14], p. 226).

A.2: The Hessian matrices $\nabla^2 f$, $\nabla^2 h_i$, where $h = (h_1,\ldots,h_m)'$ are Lipschitz continuous within an open neighborhood around $\bar{x}$.

Now let S be an arbitrary bounded subset of R^m . Consider also for any scalar $c > 0$ and any vector $y \in S$ the <u>augmented Lagrangian function</u>

$$L(x,y,c) = f(x) + y'h(x) + \frac{c}{2} \|h(x)\|^2 \tag{2}$$

where prime denotes transposition and $\|\cdot\|$ is the usual Euclidean norm. We shall be interested in algorithms of the following general (and imprecise) form. Such algorithms may be viewed as generalized quadratic penalty function methods.

<u>Step 1</u>: Given $c_k > 0$, $y_k \in S$ find a (perhaps approximate) minimizing point x_k of the function $L(x,y_k,c_k)$ defined by (2) .

<u>Step 2</u>: Determine $c_{k+1} > 0$, $y_{k+1} \in S$ on the basis of x_k,y_k,c_k according to some procedure and return to step 1.

It is easy to verify that as $c \to \infty$ we have $L(x,y,c) \to \infty$ for all $y \in S$ and all infeasible vectors x . It is thus evident that one may devise a penalty function method based on sequential unconstrained minimization of $L(x,y_k,c_k)$ for any sequences $c_k \to \infty$, $\{y_k\} \subset S$. This method exhibits the same type of convergence properties as the usual quadratic penalty method [7] which is in fact obtained by taking $y_k = 0$ for all k . The question which is most interesting, however, is to determine methods of updating y_k which result in desirable behavior such as accelerated convergence. Before proceeding to a detailed analysis, let us consider a heuristic argument which shows that <u>it is advantageous to select y_k, as close as possible to the Lagrange multiplier $\bar{y}$</u>.

Let p be the <u>primal functional</u> or perturbation function corresponding to problem (1)

$$p(u) = \min_{h(x) = u} f(x).$$

In the above equation the minimization is understood to be local within an appropriate neighborhood of $\bar{x}$. Also p is defined locally on a neighborhood of $u = 0$. It is known that

$$p(0) = f(\bar{x}) = \text{optimal value of problem (1)}$$
$$\nabla p(0) = -\bar{y}.$$

We can write

$$\min_x L(x,y,c) = \min_u \min_{h(x)=u} f(x) + y'h(x) + \frac{c}{2}||h(x)||^2 =$$

$$= \min_u \{p(u) + y'u + \frac{c}{2}||u||^2\}.$$

Now if one assumes that $p(u)$ is twice differentiable in a neighborhood of zero then it follows that for c sufficiently large the function within braces above is convex in that neighborhood and the necessary condition for optimality

$$\nabla p(u) + y + cu = 0$$

is also a sufficient condition for optimality. Hence the condition $\nabla p(0) + \bar{y} = 0$ implies that $\bar{u} = 0$ minimizes $p(u) + \bar{y}'u + \frac{c}{2}\|u\|^2$. Applying the implicit function theorem in the above equation at $\{u = 0, y = \bar{y}\}$ and denoting by $u(y)$ a local solution for u in terms of y we obtain $\lim_{y\to\bar{y}} u(y) = 0$ and

$$\lim_{y\to\bar{y}} \min_x L(x\ y,c) = p(0) = \text{optimal value of problem (1)}.$$

In other words, for c large enough, values of y close to $\bar{y}$ lead to minimum values of the augmented Lagrangian $L(x,y,c)$ close to the optimal value of the problem.

The preceding argument leads us to the conclusion that the convergence of the generalized penalty algorithm could be enhanced if, at the k-th minimization, a vector y_k close to $\bar{y}$, were to be used in the augmented Lagrangian (2) in place of $y_k = 0$ which is used in the ordinary penalty function method. But such vectors are readily available. It is easily shown that if $x(y,c)$ min-minimizes $L(x,y,c)$ then the vector

$$\tilde{y} = y + c\ h[x(y,c)]$$

is an approximation to the Lagrange multiplier $\bar{y}$ in the sense that $\lim_{c\to\infty} \tilde{y} = \bar{y}$. Thus the preceding observation suggests the following algorithm (in rough form) which is in fact identical to the method of multipliers as proposed by Hestenes and Powell.

Step 1: Given y_k, $c_k > 0$, let x_k be a minimizing point of $L(x, y_k, c_k)$.

Step 2: Update y_k be setting $y_{k+1} = y_k + c_k h(x_k)$. Update c_k according to some scheme and return to step 1.

The conclusion stated earlier can be rigorously established by making use of the following proposition which serves both as a global convergence result as well as a rate of convergence result for the method of multipliers. The proof can be found in [3].

Proposition 1: There exists a scalar $c^* \geq 0$ such that for every $c > c^*$, $y \in S$ the augmented Lagrangian $L(x,y,c)$ of (2) has a unique minimizing point $x(y,c)$ over some open ball centered at $\bar{x}$. Furthermore, for some scalar $M > 0$ we have

$$(3) \qquad \|x(y,c) - \bar{x}\| \leq \frac{M\|y - \bar{y}\|}{c} \quad \forall\, c > c^*,\ y \in S$$

$$(4) \qquad \|\tilde{y}(y,c) - \bar{y}\| \leq \frac{M\|y - \bar{y}\|}{c} \quad \forall\, c > c^*,\ y \in S$$

where the vector $\tilde{y}(y,c) \in R^m$, is given by

$$(5) \qquad \tilde{y}(y,c) = y + c\, h[x(y,c)].$$

Some important conclusions can now be obtained from the result of Proposition 1. Assuming that

$0 \in S$ we have that in the quadratic penalty method ($y_k = 0$) we obtain convergence if $c_k \to \infty$ and, furthermore, the sequences $\{x(0,c_k)\}$, $\{\tilde{y}(0,c_k)\}$ converge to $\bar{x}, \bar{y}$ respectively at least as fast as $\frac{M\|\bar{y}\|}{c_k}$. It is evident, however, from the proposition that a great deal can be gained if the vectors y_k are not held fixed but rather are updated by means of the iteration of the multiplier method

$$y_{k+1} = \tilde{y}(y_k, c_k) = y_k + c_k h[x(y_k, c_k)]. \tag{6}$$

In order to guarantee that the sequence $\{y_k\}$ remains bounded we require that updating takes place provided the resulting vector y_{k+1} belongs to the set S. Otherwise, y_k is left unchanged or perhaps projected on S. Of course, the choice of S is arbitrary and in particular we may assume that S contains $\bar{y}$ as an interior point. Under these circumstances we have that if $c_k \to \infty$ then

$$\lim_{k\to\infty} \frac{\|y_{k+1} - \bar{y}\|}{\|y_k - \bar{y}\|} = 0 \tag{7}$$

i.e. the sequence $\{y_k\}$ converges to $\bar{y}$ superlinearly. If $c_k \to c < \infty$ where c is sufficiently large, then

$$\limsup_{k\to\infty} \frac{\|y_{k+1} - \bar{y}\|}{\|y_k - \bar{y}\|} \le \frac{M}{c} \tag{8}$$

i.e. $\{y_k\}$ converges to $\bar{y}$ at least linearly with convergence ratio inversely proportional to c .

In conclusion, the method of multipliers defined by (6) converges from an arbitrary starting point within the bounded set S provided c_k is sufficiently large after some index $\bar{k}$ and the unconstrained minimizations yield the points $x(y_k, c_k)$ for all $k \geq \bar{k}$. This convergence result is of a <u>global</u> nature and is substantially stronger than local convergence results for the method of multipliers, first obtained by Buys [6] and Rupp [25]. Such local results assume that the initial choice of multiplier be sufficiently close to $\bar{y}$ in order to show convergence. It should be noted, of course, that our result also includes an assumption of a local nature namely that the unconstrained minimizations yield after a certain index the local minima $x(y_k, c_k)$ which are closest to the same local minimum $\bar{x}$ - a fact that often cannot be guaranteed a priori in the presence of other local minima. Nonetheless this restriction does not appear to be very severe since it is usually the case that penalty methods "lock into" one and the same local minimum of the problem. This is particularly so in view of the usual practice of starting each new unconstrained minimization at the final point of the previous one. The proposition also serves to demonstrate that the multiplier method offers distinct advantages over the quadratic penalty method in that it avoids the necessity of increasing c_k to infinity, and furthermore, the estimate of its convergence rate is much more favorable. For example, if $c_k = s^k$, $s > 1$ then for the penalty method we have

$$\|x(0,c_k) - \bar{x}\| \le M \|\bar{y}\| s^{-k}$$

while in the multiplier method with $y_0 = 0$

$$\|x(y_k,c_k) - \bar{x}\| \le M^{k+1} \|\bar{y}\| s^{-(1+2+\cdots+k)}$$

The ratio of the two bounds in the above inequalities is $\prod_{i=0}^{k-1} s^i M^{-1}$ and tends to infinity as $k \to \infty$.

We turn now our attention to a generalized penalty method where, given c_k, y_k the augmented Lagrangian $L(x,y_k,c_k)$ of (2) is not minimized exactly but rather the minimization process is terminated when a certain stopping criterion is satisfied. We consider two different stopping criteria. According to the first criterion, minimization of $L(x,y_k,c_k)$ is terminated at a point x_k satisfying

$$\|\nabla L(x_k,y_k,c_k)\| \le \frac{\gamma_k}{c_k} \tag{9}$$

where $\{\gamma_k\}$ is a bounded sequence with $\gamma_k \ge 0$. According to the second criterion minimization is terminated at a point x_k satisfying

$$\|\nabla L(x_k,y_k,c_k)\| \le \min \left\{\frac{\gamma_k}{c_k}, \gamma_k' \|h(x_k)\| \right\} \tag{10}$$

where $\{\gamma_k\}$, $\{\gamma'_k\}$, are bounded sequences with $\gamma_k \geq 0$, $\gamma'_k \geq 0$.

It is shown in [3] that when the criterion (9) is used to terminate the unconstrained minimization of the augmented Lagrangian the estimates (3), (4) for the multiplier method (6), take the form

$$\|x_k - \bar{x}\| \leq \frac{M(\|y_k - \bar{y}\| + \gamma_k^2)^{1/2}}{c_k} \tag{11}$$

$$\|y_{k+1} - \bar{y}\| \leq \frac{M(\|y_k - \bar{y}\| + \gamma_k^2)^{1/2}}{c_k} \tag{12}$$

where $M > 0$ is some constant and c_k is sufficiently large. When the criterion (10) is employed we have

$$\|x_k - \bar{x}\| \leq \frac{M(4\gamma_k'^2 + 1)^{1/2} \|y_k - \bar{y}\|}{c_k} \tag{13}$$

$$\|y_{k+1} - \bar{y}\| \leq \frac{M(4\gamma_k'^2 + 1)^{1/2} \|y_k - \bar{y}\|}{c_k} \tag{14}$$

The above estimate can be used to show convergence of any sequence $\{x_k, y_k\}$ generated by the iterations (6) and the termination criterion (10), provided c_k is sufficiently large after a certain index and $\bar{y}$ is an interior point of S . Furthermore, y_k converges to $\bar{y}$ at least linearly when $c_k \to c < \infty$ and superlinearly when $c_k \to \infty$. However, for the termination criterion (9) linear convergence cannot be guaranteed and in fact an example given in [2] shows that convergence may not be linear. In addition for this termination criterion it is necessary to increase c_k to infinity in order to achieve global convergence. This latter restriction, however, may be removed by using a sequence $\{\gamma_k\}$ converging to zero.

It should be noted that the employment of inexact minimization in the method of multipliers is of both computational and theoretical significance. From the computational point of view inexact minimization usually results in significant computational savings. From the theoretical point of view inexact minimization coupled with the termination criterion (10) yields a procedure which is similar to primal-dual methods of the Lagrangian type [1], [16] while it guarantees global convergence of the sequences of primal and dual variables generated. In Lagrangian methods usually local convexity assumptions are required in order to guarantee merely local convergence. It is worth noting that the utilization of a penalty function to convexify locally the problem and thereby ensure local convergence of a Lagrangian method has been pointed out as early as 1958 by Arrow and Solow [1].

III. A Global Duality Framework for the Method of Multipliers

In this section we utilize the results of section 2 to construct a duality framework for problem (1). In contrast with past formulations for nonconvex problems (see e.g. [6], [14]) the framework is <u>global</u> in nature (at least in as much as the dual variables are concerned). By this we mean that the dual functional is an everywhere defined real valued concave function. The theory is similar in spirit with the one recently proposed by Rockafellar [22] under weaker assumptions, and the one of Buys [6] which is local in nature. Our construction however is more suitable to the analysis of algorithms since in our case <u>the dual functional has strong differentiability properties</u>. Furthermore its value and derivatives within an arbitrary open bounded set may be computed by local unconstrained minimization of the augmented Lagrangian similarly as for convex problems. In this way the iteration of the multiplier method can be interpreted as a gradient iteration in a global sense.

For any vector $u \in R^m$ consider the minimization problem

$$(15) \qquad \min_{h(x) = u} f(x)$$

Now by applying the implicit function theorem to the system of equations,

$$\nabla f(x) + \sum_{i=1}^{m} y^i \nabla h_i(x) = 0, \ h_i(x) = u_i \ , \ i=1,\dots,m$$

and using assumption A.1 we have the following lemma:

<u>Lemma</u>: Under assumption A.1 there exist positive scalars β and δ such that for every u with $\|u\| < \beta$ problem (15) has a unique solution $x(u)$

within the open ball $B(\bar{x},\delta)$ with a Lagrange multiplier $y(u)$ satisfying $\|y(u) - \bar{y}\| < \delta$. Furthermore, the functions $x(u)$, $y(u)$ are continuously differentiable within $B(0,\beta)$ and satisfy $x(0) = \bar{x}$, $y(0) = \bar{y}$.

We define now the <u>primal functional</u> $p: B(0,\beta) \to R$ by means of

$$p(u) = \min_{\substack{h(x)=u \\ x\in B(\bar{x},\delta)}} f(x) = f[x(u)]. \tag{16}$$

It follows from the implicit function theorem that

$$\nabla p(u) = -y(u), \quad u\in B(0,\beta) \tag{17}$$

and, since $y(u)$ is continuously differentiable, we have that p is twice continuously differentiable on $B(0,\beta)$. Without loss of generality we assume that the Hessian matrix of p is uniformly bounded on $B(0,\beta)$.

Now for any $c \geq 0$ consider the function

$$p_c(u) = p(u) + \frac{c}{2}\|u\|^2. \tag{18}$$

It is clear that there exists a constant $\mu > 0$ such that for all $c \geq \mu$ the Hessian matrix of p_c is positive definite on $B(0,\beta)$ and hence p_c is strictly convex on $B(0,\beta)$. We define for every $c \geq \mu$ the <u>dual functional</u> $d_c: R^m \to R$ by means of

$$d_c(y) = \inf_{u\in B(0,\beta)} \{p(u) + \frac{c}{2}\|u\|^2 + y'u\} = \inf_{u\in B(0,\beta)} \{p_c(u)+y'u\}. \tag{19}$$

We note that this way of defining the dual functional is not unusual since it corresponds to a perturbation function taking the value $p_c(u)$ on $B(0,\beta)$ and $+\infty$ outside $B(0,\beta)$.

Under assumption A.1 the function d_c of (19) has the following easily proved properties which we state as a proposition:

Proposition 2: a) The function d_c is a real valued, everywhere continuously differentiable concave function. Furthermore it is twice continuously differentiable on the open set $A = \{y | y = -\nabla p_c(u),\ u \in B(0,\beta)\}$. b) For any $y \in A$ the infimum in (19) is attained at a unique point $u_y \in B(0,\beta)$ and we have $\nabla d_c(y) = u_y$, $\nabla^2 d_c(y) = -[\nabla^2 p_c(u_y)]^{-1}$. c) The function d_c has a unique maximizing point, the Lagrange multiplier $\bar{y}$.

We now proceed to show that the value and the derivatives of the dual functional d_c can be obtained by local minimization of the augmented Lagrangian $L(x\ y,c)$ of (2) provided c is sufficiently large. Let S be any open bounded subset of R^m. Then for any $y \in S$, by Proposition 1, we have that for c sufficiently large.

$$\|x(y,c) - \bar{x}\| \leq \frac{M\|y - \bar{y}\|}{c} < \delta ,$$

$$\|\tilde{y}(y,c) - \bar{y}\| \leq \frac{M\|y - \bar{y}\|}{c} < \delta,$$

$$\|\tilde{u}\| < \beta$$

where

$$\tilde{y}(y,c) = y + c\ h[x(y,c)], \quad \tilde{u} = h[x(y,c)].$$

Furthermore we have

$$\nabla f[x(y,c)] + \tilde{y}(y,c)\ \nabla h[x(y,c)] = 0\ .$$

It follows from the implicit function theorem and the lemma that $x(y,c)$ is the unique minimizing point in problem (15) when $u = \tilde{u}$. This implies

$$p(\tilde{u}) = f[x(y,c)],\ \nabla p(\tilde{u}) = -\ \tilde{y}(y,c) = -y - c\tilde{u}$$

and therefore

$$\nabla p_c(\tilde{u}) + y = 0\ .$$

Hence $\tilde{u}$ attains the infimum in the right hand side of (19) and by part b) of Proposition 2

$$\nabla d_c(y) = \tilde{u} = h[x(y,c)]$$

$$\nabla^2 d_c(y) = -[\nabla^2 p_c(\tilde{u})]^{-1}\ .$$

Furthermore

$$d_c(y) = p(\tilde{u}) + y'\tilde{u} + \frac{c}{2}\ ||\tilde{u}||^2$$

$$= f[x(y,c)] + y'h[x(y,c)] + \frac{c}{2}\ ||h[x(y,c)]\ ||^2$$

$$= \min_x\ L(x,y,c)$$

where the minimization above is understood to be local in the sense of Proposition 1. In addition a straightforward calculation [6], [14] yields

$$(20)\quad D_c(y) = \nabla^2 d_c(y) =$$

$$-\nabla h[x(y,c)]\{\nabla^2 L[x(y,c),y,c]\}^{-1}\nabla h[x(y,c)]'$$

where $\nabla h[x(y,c)]$ is the $m \times n$ matrix having as rows the gradients $\nabla h_i[x(y,c)]$, $i = 1,\ldots,m$ and $\nabla^2 L$ denotes the Hessian matrix of the augmented Lagrangian L with respect to x. Thus we have proved the following proposition.

Proposition 3: Let S by any open bounded subset of R^m and let assumptions A.1, A.2 hold. Then there exists a scalar $c^* \geq 0$ such that for every $y \in S$ and every $c > c^*$ the dual functional d_c satisfies

$$d_c(y) = f[x(y,c)] + y'h[x(y,c)] + \frac{c}{2}\|h[x(y,c)]\|^2 =$$

$$= \min_x L(x\ y,c)$$

$$\nabla d_c(y) = h[x(y,c)]$$

where $x(y,c)$ is as in Proposition 1. Furthermore d_c is twice continuously differentiable on S and $\nabla^2 d_c(y)$ is given by (20).

It is now clear that the iteration of the method of multipliers can be written for c sufficiently large

$$y_{k+1} = y_k + c\ \nabla d_c(y_k)$$

and hence can be viewed as a fixed stepsize gradient iteration for maximizing the dual functional d_c.

Thus one may obtain a tight rate of convergence result by utilizing a known result on gradient methods [17]. This result however is rather uniformative since it involves the eigenvalues of the matrix D_c of (20) which strongly depend on c. A modified version of this result which is more amenable to proper interpretation is given in [2] together with an analysis of the convergence rate aspects of the method of multipliers in the presence of inexact minimization.

The primal-dual interpretation of the multiplier method suggests also several possibilities for modification of the basic iteration. One such modification was suggested in [2], [4]. Another interesting possibility rests on the fact that when second derivatives are calculated during the unconstrained minimization cycle, then one obtains the Hessian matrix D_c of (20) in addition to the gradient ∇d_c. Thus it is possible to carry out a Newton iteration aimed at maximizing d_c in place of the gradient iteration corresponding to the method of multipliers. It is also possible to use a variable metric method for maximization of d_c. Such possibilities have already been suggested by Buys [6] who in addition provided some local convergence results. It is to be noted however that for large scale problems arising for example in optimal control, where the number of primal and dual variables may easily reach several hundreds or even thousands, such modifications do not seem to be attractive. This is particularly so since the simple gradient iteration already has excellent convergence rate. It is also interesting to observe the the global duality framework may be used to show that the pair $(\bar{x},\bar{y})$ is an unconstrained saddle point of the augmented Lagrangian L for c

sufficiently large, where unconstrained minimization with respect to x is local. This fact holds even if inequality constraints are present, i.e. the usual nonnegativity constraints on the dual variables need not be taken into account. In this way the application of Lagrangian methods for solving the saddle point problem is considerably enhanced. This observation was first made by Rockafellar [19] and was further exploited by Mangasarian [15].

IV. Inequality Constraints and Convex Programming

As pointed out in the introduction inequality constraints may be treated in a simple way by introducing slack variables. Indeed the problem

$$\min_{\substack{g_j(x)\le 0 \\ j=1,\dots,r}} f(x) \tag{21}$$

is equivalent to the equality constrained problem

$$\min_{\substack{g_j(x)+z_j^2 = 0 \\ j=1,\dots,r}} f(x) \tag{22}$$

where $z_1,\dots,z_r$ represent additional variables. Now if $\bar{x}$ is an optimal solution of problem (21) satisfying the second order sufficiency conditions (including strict complementarity), then $(\bar{x},|g_1(\bar{x})|^{\frac{1}{2}},\dots,|g_r(\bar{x})|^{\frac{1}{2}})$ is an optimal solution for problem (22) satisfying the second order sufficiency conditions for optimality and hence it is covered by the theory of Sections 2 and 3. Thus one may use the multiplier method for solving problem (22) instead of problem (21). On the other hand slack variables need not be present explicitly in the computations since the minimization of the augmented Lagrangian

$$L(x,z,y,c) = f(x) + \sum_{j=1}^{r} y^j [g_j(x) + z_j^2] + \frac{c}{2} \sum_{j=1}^{r} [g_j(x) + z_j^2]^2$$

can be carried out first with respect to $z_1, \ldots, z_r$ yielding

$$\tilde{L}(x,y,c) = \min_z L(x,z,y,c) =$$

$$= f(x) + \frac{1}{2c} \{ \sum_{j=1}^{r} [\max(0, y^j + cg_j(x)]^2 - (y^j)^2 \}.$$

The optimal values of z_j are given in terms of x,y,c by

$$(23) \quad z_j^2(x,y,c) = \max[0, -\frac{y^j}{c} - g_j(x)], \ j = 1, \ldots, r .$$

Now minimization of $\tilde{L}(x,y,c)$ with respect to x yields a vector $x(y,c)$ and the multiplier method iteration in view of (23) takes the form

$$(24) \quad y_{k+1}^j = y_k^j + c[g_j[x(y,c)] + z_j^2[x(y,c),y,c]] =$$

$$= \max [0, y_k^j + cg_j[x(y,c)]], \ j=1,\ldots,r.$$

Thus there is no difference in treating equality or inequality constraints at least within the second order sufficiency assumption framework of this paper. It is worth noting that if the sequences

$\{x_k\}$, $\{y_k\}$ generated by the multiplier method converge to $\bar{x}, \bar{y}$ respectively then one may easily see from (24) that the approximate Lagrange multipliers which correspond to inactive constraints converge to zero within a finite number of iterations.

We note that when additional structure such as convexity is inherently present in the problem then one can considerably weaken the assumptions of section 2 and 3 while obtaining much more powerful convergence and duality results. We refer to the papers by Rockafellar [20], [21] and Kort and the author [10] - [13], [4] for an exhaustive analysis of multiplier methods for convex programming under very weak assumptions. An important characteristic of the method of multipliers when applied to convex programming problems is that global convergence for both exact and approximate minimization is achieved for <u>any</u> positive value of the penalty parameter thus completely eliminating the ill-conditioning problem. This is not really very surprising since the primary role of the penalty parameter is to induce a convexification effect which in convex programming problems is already present.

Finally we mention that the method of multipliers shares with the quadratic penalty method one weakness. Whenever the objective function decreases at a rate higher than quadratic as $||x|| \to \infty$, the infimal value of the augmented Lagrangian may be $-\infty$ thereby introducing serious computational difficulties. A typical example is the simple one-dimensional problem $\min_{x=0} \{-|x|^3\}$. One way to bypass the difficulty is to introduce penalty functions with order of growth higher than quadratic or even barrier functions in place of the quadratic penalty function. For example one may consider, in place of the quadratic penalty function t^2, penalty functions of the form

$$(25) \quad p(t) = |t|^{\rho_1} + |t|^{\rho_2} + \cdots + |t|^{\rho_s} \quad \rho_1, \ldots, \rho_s > 1.$$

For analysis related to such more general multiplier methods see the paper by Mangasarian [15] and the papers by Kort and the author [10] - [13]. An interesting fact related to such multiplier methods is that the rate of convergence strongly depends on the choice of the penalty function. Thus for the case of a convex programming problem it is shown in [11] - [13] under mild assumptions that one may obtain superlinear convergence rate with arbitrarily high Q-order of convergence by proper choice of penalty function. This convergence rate result for a penalty function such as (25) and for a dual functional satisfying a certain quadratic growth condition has the form

$$\lim_{k\to\infty} \sup \frac{\|y_{k+1} - \bar{y}\|}{\|y_k - \bar{y}\|^{\alpha}} \le K < \infty$$

where

$$\alpha = \max \{1, \frac{1}{\rho - 1}\}$$

and

$$\rho = \min \{\rho_1, \ldots, \rho_s\} .$$

Thus for $1 < \rho < 2$ superlinear convergence is obtained. This improvement in convergence rate is associated however with certain ill-conditioning effects since when $1 < \rho < 2$ the penalty function (25) is not twice differentiable at $t = 0$.

References

[1] Arrow, K. J., Hurwicz, L., and Uzawa, H., (1958) "Studies in Linear and Nonlinear Programming," Stanford University Press, Stanford, Calif.

[2] Bertsekas, D. P., (Jan 1973) "Combined Primal Dual and Penalty Methods for Constrained Minimization," EES Dept. Working Paper, Stanford University, Stanford, Calif., to appear in SIAM J. on Control, Vol. 13, No. 3, Aug. 1975.

[3] Bertsekas, D. P., (August 1973) "On Penalty and Multiplier Methods for Constrained Minimization," EES Dept. Working Paper, Stanford University, Stanford, Calif., to appear in SIAM J. on Control.

[4] Bertsekas, D. P., (August 1973) "On the Method of Multipliers for Convex Programming," EES Dept. Working Paper, Stanford University, Stanford, Calif., submitted for publication.

[5] Bertsekas, D. P., (Dec. 1973) "Convergence Rate of Penalty and Multiplier Methods," Proceedings of 1973 IEEE Conference on Decision and Control, San Diego, Calif., pp. 260-264.

[6] Buys, J. D., (June 1972) "Dual Algorithms for Constrained Optimization," Ph.D. Thesis Rijksuniversiteit de Leiden.

[7] Fiacco, A. V., and McCormick, G. P., (1968) "Nonlinear Programming: Sequential Unconstrained Minimization Techniques," J. Wiley, New York, N. Y., 1968.

[8] Haarhoff, P. C., and Buys, J. D., (1970) "A New Method for the Optimization of a Nonlinear Function Subject to Nonlinear Constraints," Computer Journal, Vol. 13, pp. 178-184.

[9] Hestenes, M. R., (1969) "Multiplier and Gradient Methods," Journal of Optimization Theory and Applications, Vol. 4, No. 5, pp. 303-320.

[10] Kort, B. W., and Bertsekas, D. P., (Dec. 1972) "A New Penalty Function Method for Constrained Minimization," Proc. of 1972 IEEE Decision and Control Conference, New Orleans, La.

[11] Kort, B. W., and Bertsekas, D. P., (August 1973) "Combined Primal-Dual and Penalty Methods for Convex Programming," EES Dept. Working Paper, Stanford University, Stanford, Calif., submitted for publication.

[12] Kort, B. W., and Bertsekas, D. P., (Dec. 1973) "Multiplier Methods for Convex Programming," Proceedings of 1973 IEEE Conference on Decision and Control, San Diego, Calif., pp. 428-432.

[13] Kort, B. W., "Combined Primal-Dual and Penalty Function Algorithms for Nonlinear Programming," Ph.D. Dissertation, Dept. of Electrical Engineering, Stanford University, Stanford, Calif., forthcoming.

[14] Luenberger, D. G., (1973) "Introduction to Linear and Nonlinear Programming," Addison-Wesley, Inc.

[15] Mangasarian, O. L., (1974) "Unconstrained Lagrangians in Nonlinear Programming," Computer Science Tech. Report #201, Univ. of Wisconsin, Madison.

[16] Miele, A., Mosley, P. E., Levy, A. V., and Coggins, G. M., (1972) "On the Method of Multipliers for Mathematical Programming Problems," Journal of Optimization Theory and Applications, Vol. 10, No. 1, pp. 1-33.

[17] Polyak, B. T., (1963) "Gradient Methods for the Minimization of Functionals," Zh. Vychisl. Mat. Mat. Fiz., Vol. 3, No. 4., pp. 643-653.

[18] Powell, M. J. D., (1969) "A Method for Nonlinear Constraints in Minimization Problems," in Optimization, R. Fletcher (ed.), Academic Press, New York, pp. 283-298.

[19] Rockafellar, R. T., (1971) "New Applications of Duality in Convex Programming," written version of talk at 7th International Symposium on Math. Programming (the Hague, 1970) and elsewhere, published in the Proc. of the 4th Conference on Probability, Brasov, Romania.

[20] Rockafellar, R. T., (1973) "A Dual Approach to Solving Nonlinear Programming Problems by Unconstrained Optimization," Math. Prog. Vol. 5, pp. 354-373.

[21] Rockafellar, R. T., (1973) "The Multiplier Method of Hestenes and Powell Applied to Convex Programming," J. Opt. Theory Appl., Vol. 12, No. 6.

[22] Rockafellar, R. T., (1974) "Augmented Lagrange Multiplier Functions and Duality in Nonconvex Programming," SIAM J. Control, Vol. 12, No. 2.

[23] Rockafellar, R. T., (1974) "Penalty Methods and Augmented Lagrangians in Nonlinear Programming," Proceedings of the 5th IFIP Conference on Optimization Techniques, Rome, 1973, Springer-Verlag.

[24] Rockafellar, R. T., (to appear) "Solving a Nonlinear Programming Problem by way of a Dual Problem," Symposia Matematica.

[25] Rupp, R. D., (1973) "A Nonlinear Optimal Control Minimization Technique," Trans. of the Amer. Math. Soc., 178, pp. 357-381.

RATE OF CONVERGENCE OF THE METHOD OF MULTIPLIERS WITH INEXACT MINIMIZATION

by

Barry W. Kort[1)]

ABSTRACT

The Method of Multipliers is a primal-dual algorithm based on sequential unconstrained minimization of a generalized Lagrangian. Characteristic to the method is a dual iteration in which the current estimate of the Lagrange multiplier is updated at the end of each unconstrained minimization. The method may be defined for a large class of generalized Lagrangians. When applied to a convex programming problem, the method is globally convergent to an optimal solution - Lagrange multiplier pair.

The unconstrained minimizations need not be carried out exactly. A computationally implementable stopping rule is given for terminating the unconstrained minimizations short of locating the exact minimum. The stopping rule is shown to preserve the global convergence of the algorithm.

The rate of convergence is derived for both exact and inexact minimization and is shown to depend primarily on the type of penalty function used in constructing the generalized Lagrangian. Although inexact minimization may worsen the rate (and order) of convergence, it is possible to operate the

[1]Bell Laboratories, Holmdel, New Jersey 07733

stopping rule in such a way as to maintain the rate of convergence obtainable with exact minimization.

The method is explained in terms of geometric interpretations which illustrate the mechanism by which the algorithm locates the Lagrange multiplier. The figures also show why the method is superior to ordinary penalty techniques.

1. Introduction

The Method of Multipliers was proposed as early as 1968 by Hestenes [2] and Powell [6]. The method utilizes an augmented Lagrangian function which was previously studied by Arrow and Solow [1]. Since 1968, no fewer than 16 researchers have contributed to the burgeoning literature on the subject. Extensive references may be found in [4] and [5].

This paper summarizes the main results of research jointly conducted by Professor Dimitri Bertsekas and the author. Since the details of that work are exhaustively covered in [4] and [5], no attempt will be made here to repeat that analysis. The present objective is simply to describe the method, review its properties, and provide some heuristic insight into the nature of the method. The present discussion focuses on the method as applied to convex programs.

2. The Method of Multipliers

The method to be described is a combined primal-dual and penalty function algorithm for solving the non-linear programming problem. One chooses a modified or penalized Lagrangian function by selecting a penalty function from a large class. The penalized Lagrangian incorporates a dual variable (Lagrange multiplier) which has the same interpretation as the classical Lagrange multiplier. Fixing the dual variable at some estimate of the Lagrange multiplier, one may minimize the penalized Lagrangian with respect to the decision variable. Upon completion of the unconstrained minimization, the dual variable may be updated via a simple formula, to yield a better estimate of the Lagrange multiplier. The Method of Multipliers simply iterates this process. Thus the method consists of solving a sequence of unconstrained minimizations of the penalized Lagrangian with a dual iteration at the end of each minimization.

To make the algorithm more attractive from a practical point of view, one would like to require only approximate minimization. Consequently we incorporate a stopping criterion for inexact minimization. This criterion has the desirable property that it does not destroy the convergence properties exhibited by the algorithm under exact minimization.

3. Notation

The convex programming problem (CPP) is stated as

$$\begin{aligned} &\text{minimize} && f_0(x) && \\ (1)\quad &\text{subject to} && f_i(x) \le 0 && i = 1,\dots,m \\ & && f_i(x) = 0 && i = m+1,\dots,d \end{aligned}$$

where the f_i $i = 0,\dots,m$ are extended real valued closed proper convex functions on R^n and f_i,

$i = m+1,\ldots,d$ are affine functions on R^n. Note that there are no differentiability assumptions on the convex functions. Set constraints of the form $x \in X \subset R^n$ are assumed to be incorporated into the objective function f_0 by defining $f_0(x) = +\infty$, $x \notin X$.

We assume that the CPP satisfies the following assumptions:

A.i) $\text{dom } f_0 \subset \text{dom } f_i \quad i = 1,\ldots,m$
$\text{ri dom } f_0 \subset \text{ri dom } f_i \quad i = 1,\ldots,m;$

A.ii) the CPP possesses a non-empty and bounded solution set $X^* \subset R^n$;

A.iii) the set of Lagrange multipliers $Y^* \subset R^d$ associated with the CPP is non-empty and bounded.

The ordinary Lagrangian is denoted

$$L(x;y) = \begin{cases} f_0(x) + \Sigma\, y_i f_i(x) & y \in Y \\ -\infty & y \notin Y \end{cases} \tag{2}$$

where $Y = R_+^m \times R^{d-m}$. The penalized Lagrangian is constructed using a convex penalty function $\phi: R \to [0,\infty]$ and a convex-concave saddle function $p: R^2 \to [-\infty,\infty]$. The functions ϕ and p may be selected from the classes determined by the following specifications.

The penalty function $\phi: R \to [0,\infty]$ must satisfy the following six properties for some choice of $b_1 \in [-\infty,0)$ and $b_2 \in (0,\infty]$:

a) ϕ is continuous and has continuous derivative on (b_1,b_2).

b) ϕ is strictly convex and closed.

c) $\phi(0) = 0$

d) $\nabla\phi(0) = 0$

e) $\lim_{t \downarrow b_1} \nabla\phi(t) = -\infty$

f) $\lim_{t \uparrow b_2} \nabla\phi(t) = +\infty$.

Examples: $\phi(t) = \frac{1}{2}t^2$.

$\phi(t) = \cosh(t) - 1$.

The saddle function $p:R^2 \to [-\infty,\infty]$ must satisfy the following eight properties for some choice of $b \in (0,\infty]$:

a) p is continuous relative to $(-\infty,b) \times R_+$ and possesses both partial derivatives on $(-\infty,b) \times R_+$ including the right partial derivative

$$\nabla_2 p(t;y)\Big|_{y=0} \quad \text{for all } t < b.$$

b) For each fixed $y \in R$, $p(\cdot;y)$ is closed and convex on R with the following strict convexity requirement:

if i) $y \geq 0$ and $t_0 < b$

and ii) $t_0 > 0$ or $\nabla_1 p(t_0;y) > 0$ then

$$p(t;y) - p(t_0;y) > (t-t_0)\nabla_1 p(t_0;y) \quad t \neq t_0 .$$

c) $p(t;\cdot)$ is concave on R for each fixed $t \in R$.

For all $y \geq 0$,

d) $p(0;y) = 0$

e) $\nabla_1 p(0;y) = y$

f) $\lim_{t \to -\infty} \nabla_1 p(t;y) = 0$

g) $\lim_{t \uparrow b} \nabla_1 p(t;y) = +\infty$

h) $\inf_t p(t;y) > -\infty$.

Example: $$p(t;y) = \begin{cases} yt + \frac{1}{2}t^2 & t \geq -y \\ -\frac{1}{2}y^2 & t \leq -y. \end{cases}$$

Define the penalized Lagrangian for $r > 0$ by

$$\text{(3)} \quad L_r(x;y) = \begin{cases} f_0(x) + \sum_{i=1}^{m} rp[f_i(x)/r;y_i] + \\ \qquad \sum_{i=m+1}^{d} y_i f_i(x) + r\phi[f_i(x)/r] \\ \qquad \text{if } y_i \in \text{dom}_2 p \quad i = 1,\ldots,m \\ -\infty \qquad \text{otherwise.} \end{cases}$$

The scalar $r > 0$ is a penalty parameter which controls the severity of the penalty. In particular note that (for $y \varepsilon Y$)

$$\lim_{r\to\infty} L_r(x;y) = L(x;y)$$

$$\lim_{r\to 0} L_r(x;y) = \begin{cases} f_0(x) & \text{if } x \text{ is feasible} \\ -\infty & \text{if } x \text{ is not feasible.} \end{cases}$$

Denote by g and g_r the ordinary and penalized dual functionals

$$g(y) = \inf_x L(x;y)$$

$$g_r(y) = \inf_x L_r(x;y) \ .$$

4. Duality

The dual functionals g and g_r are concave in the dual variable y. One seeks a Lagrange multiplier y^* which maximizes the dual. As shown in [5], $g(\cdot)$

and $g_r(\cdot)$ have identical maximum sets. Having found a Lagrange multiplier y^*, one seeks an optimal solution x^* among the set of minimizers of $L_r(\cdot;y^*)$.

Proposition 1. Let $y^* \in R^d$ be a Lagrange multiplier for (1). Then x^* is an optimal solution if and only if x^* minimizes $L_r(\cdot;y^*)$.

Proof. See [5, prop. 4.5].

Just as for the ordinary Lagrangian L, a pair $(x^*;y^*)$ is an optimal solution-Lagrange multiplier pair if and only if it is a saddle point of the penalized Lagrangian L_r.

5. The Method of Multipliers Algorithm

The algorithm consists of a sequence of unconstrained minimizations of the penalized Lagrangian $L_r(\cdot;y^k)$ $k = 1,2,\ldots$. The dual variable y^k is held fixed during the minimization and is then updated prior to the next unconstrained minimization.

Basic Algorithm

Select penalty functions p and ϕ (according to the criteria of § 3) and a scalar $r^o > 0$. Select an initial estimate, $y^o \varepsilon Y$, of the Lagrange multiplier vector.

Step 1: Given y^k and r^k, find x^k to solve the unconstrained minimization

$$\min_{x \varepsilon R^n} L_{r^k}(x;y^k) \; . \tag{4}$$

Step 2: Using the x^k from step 1, set

$$y_i^{k+1} = \nabla_1 p[f_i(x^k)/r^k;y_i^k] \quad i = 1,\ldots,m \tag{5a}$$

$$\text{(5b)} \qquad y_i^{k+1} = y_i^k + \nabla\phi[f_i(x^k)/r^k] \quad i = m+1,\ldots,d \; .$$

Stop if $y^{k+1} = y^k$; otherwise select $r^{k+1} > 0$ and return to step 1. The sequence $\{r^k\}$ must remain bounded above.

Inexact Minimization

Denote s^k: $R^n \to R^d$ by

$$\text{(6a)} \qquad s_i^k(x) = \nabla_1 p[f_i(x)/r^k; y_i^k] \quad i = 1,\ldots,m$$

$$\text{(6b)} \qquad s_i^k(x) = y_i^k + \nabla\phi[f_i(x)/r^k] \quad i = m+1,\ldots,d \; .$$

In the basic algorithm, the dual iteration is just $y^{k+1} = s^k(x^k)$ where at stage k, x^k minimizes $L_{r^k}(\cdot;y^k)$. We shall retain the same dual iteration for inexact minimization, but x^k will now be only an approximate minimizing point. As shown in [5], the stopping rule should be designed to satisfy a test of the general form:

Find x^k to satisfy

$$\text{(7)} \qquad L_{r^k}(x^k;y^k) - \min L_{r^k}(\cdot;y^k) \le \varepsilon^k \; .$$

Unfortunately (7) is not an implementable test since it requires knowledge of the minimum value of the function being minimized. Thus we seek an implementable test which under appropriate assumptions will guarantee (7).

Denote by $\nabla_x L_r(x;y)$ the minimum norm element of the subdifferential $\partial_x L_r(x;y)$. (If L_r is in fact differentiable, then $\nabla_x L_r$ is just the ordinary gradient.) We propose the following implementable stopping rule:

Find x^k to satisfy

$$(8) \quad \| \nabla_x L_{r^k}(x^k;y^k) \|^2 \le \eta^k \{L(x^k;s^k(x^k)) - L_{r^k}(x^k;y^k)\}$$

where the non-negative scalar sequence $\{\eta^k\}$ is either predetermined or computed dynamically according to some rule.

The use of (8) in place of (7) is based on the following uniform convexity assumption:

A.iv) There exists a positive scalar μ such that for all values y^k and r^k generated by the algorithm,

$$L_{r^k}(z;y^k) - L_{r^k}(x;y^k) \ge \langle \nabla_x L_{r^k}(x;y^k), z-x \rangle + \frac{\mu}{2} \| z-x \|^2 \quad \forall\, z.$$

The above assumption is required only in the event that the algorithm is operated with inexact minimization, using the stopping rule (8). As shown in [5], the implementable rule (8) together with assumption A.iv guarantees that the required test of the form (7) will be satisfied.

6. Convergence Properties of the Algorithm

When applied to the convex programming problem (1), the algorithm of Section 5 exhibits the following properties.

- The minimization in step 1 is well posed: $L_{r^k}(\cdot;y^k)$ possesses a non-empty and compact minimum set at every stage k.
- If $\eta^k < 2\mu$ (at least for large k) then the sequence $\{y^k\}$ of dual variables (eventually) ascends the dual functionals g and g_r. In fact $\{y^k\}$ is a compact sequence converging to the set Y^* of Lagrange multipliers, i.e., $\| y^k - Y^* \| \to 0$.

The sequence $\{x^k\}$ is compact and converges to the set X^* of optimal solutions.

Thus every accumulation point of the sequence $\{(x^k;y^k)\}$ is an optimal solution-Lagrange multiplier pair. In other words the algorithm is globally convergent. These properties are proved in [5, Section 6]. Inexact minimization does not destroy global convergence provided (A.iv) holds and η^k is eventually taken smaller than 2μ.

7. Rate of Convergence

Rate of convergence is measured as the rate at which $\|y^k - Y^*\| \to 0$. Since we do not assume a unique Lagrange multiplier, we use the notion of convergence to a set in terms of the Euclidian distance from the point y^k to the convex set Y^*.

The convergence rate is linear or superlinear depending on the choice of penalty function and the penalty sequence $\{r^k\}$. We examine those penalty functions for which there is a $\rho \varepsilon (1,\infty)$ such that

$$\lim_{t\to 0} \frac{\phi(t)}{|t|^\rho} = K, \qquad 0 < K < \infty .$$

For example, if $\phi(t)$ is written as a sum of powers of $|t|$:

$$\phi(t) = a_1|t|^{\rho_1} + a_2|t|^{\rho_2} + \cdots \; a_j \geq 0, \; \rho_j \varepsilon (1,\infty),$$

then the scalar ρ is the smallest power in the series.

We further assume that the saddle function p is taken as

$$p(t;y) \triangleq \begin{cases} yt+\phi(t) & \text{if} \quad y+\nabla\phi(t) \geq 0 \\ \min_\tau\{y\tau+\phi(\tau)\} & \text{if} \quad y+\nabla\phi(t) < 0 . \end{cases}$$

(These assumptions do exclude some penalties ϕ and p delimited in Section 3, and the convergence rate analysis is so limited.)

In lieu of the usual regularity assumptions (e.g., second order sufficiency, linear independence of constraint gradients) we introduce the following much weaker regularity assumption.

A.v) There exists a positive scalar γ and a neighborhood $B(Y^*;\delta)$ such that

$$g(y)-\sup g \leq -\gamma \, ||y-Y^*||^2 \qquad y \in B(Y^*;\delta).$$

Compare A.v to the more typical regularity assumptions which generally imply that the dual functional g is twice differentiable with positive definite Hessian at a unique Lagrange multiplier y^*. The assumption A.v not only doesn't require twice differentiability, it doesn't require first differentiability or even that g be finite over the entire neighborhood $B(Y^*;\delta)$.

Proposition 2. (Order of Convergence, Exact Minimization)

Suppose the algorithm is operated with exact minimization ($\eta^k \equiv 0$, assumption A.iv not required). Then

$$\limsup_{k\to\infty} \frac{||y^{k+1}-Y^*||}{||y^k-Y^*||^a} < \infty$$

where

$$a = \begin{cases} 1 & \text{if} \quad \rho \geq 2 \\ 1/(\rho-1) & \text{if} \quad 1 < \rho \leq 2 . \end{cases}$$

Proof. See [5, prop. 8.6].

The proposition states that the order of convergence of $\{y^k\}$ depends on the power ρ for which $\phi(t) \sim |t|^\rho$ for small t. The surprising result is that one can obtain any order of convergence $a \geq 1$ by selecting a suitable penalty function.

Proposition 3. (Linear Convergence Rate, Exact Minimization)

Assume exact minimization; suppose $\rho = 2$ and $\phi(t)$ is twice differentiable at $t = 0$ with $\partial^2\phi(0)/\partial t^2 = 1$. Then

$$\limsup_{k\to\infty} \frac{\| y^{k+1}-y^* \|}{\| y^k-y^* \|} \leq \frac{\bar{r}}{\bar{r}+\gamma}$$

where $\bar{r} = \limsup r^k$.

Proof. See [5, prop. 8.7].

The proposition shows that penalty functions of the form $\phi(t) = a_2 t^2 + a_3|t|^3 + a_4 t^4 + \cdots$ $(a_2 > 0,\ a_j \geq 0,\ j = 3,4,\ldots)$ yield linear convergence rate, and the rate is superlinear if $r^k \to 0$.

Proposition 4. (Inexact Minimization)

Let $1 < \rho < 2$ and assume $\eta^k < 2\mu$ for large k . Then the order of convergence of $\{y^k\}$ is $a = \sigma/2$ where $\frac{1}{\sigma} + \frac{1}{\rho} = 1$.

Proof. See [5, prop. 8.12].

Corollary 4.1. If $\{\eta^k\}$ is replaced by the function

$$\eta^k(x) = c\| s^k(x)-y^k\|^{\sigma-2} \tag{9}$$

where $c > 0$ is arbitrary, the order of convergence increases to $a = \sigma-1 = \frac{1}{\rho-1}$.

Proof. See [5, cor. 8.12.] and discussion].

In other words, if the stopping rule (8) is operated with $\{\eta^k\}$ appropriately chosen (e.g., via (9)) the algorithm with inexact minimization exhibits the same order of convergence as the basic algorithm with exact minimization.

Proposition 5. (Linear Convergence Rate, Inexact Minimization)

Suppose $\rho = 2$ and $\partial^2\phi(0)/\partial t^2 = 1$. If $\bar{r}\,\bar{\eta} < 4\gamma\mu$ then $\{y^k\}$ converges linearly with convergence ratio

$$\limsup_{k\to\infty} \frac{\| y^{k+1} - y^* \|}{\| y^k - y^* \|} = \beta$$

where

$$\beta \le \frac{1 + \sqrt{1 + \frac{2\bar{\nu}}{1-\bar{\nu}}\left(1 + \frac{\gamma}{\bar{r}}\frac{1}{1-\bar{\nu}}\right)}}{2\left(1 + \frac{\gamma}{\bar{r}}\frac{1}{1-\bar{\nu}}\right)}, \tag{10}$$

$$\bar{r} = \limsup r^k ,$$

$$\bar{\eta} = \limsup \eta^k ,$$

and

$$\bar{\nu} = \bar{\eta}/2\mu \le 1 .$$

(If $\bar{r} = 0$, one has $\beta = 0$. If $\bar{\nu} = 1$ one has $\beta \le \sqrt{\bar{r}/2\gamma}$.)

Proof. See [5, prop. 8.13].

Corollary 5.1. If $\eta^k \to 0$, the bound (10) reduces to

$$\beta \le \frac{\bar{r}}{\bar{r}+\gamma} .$$

Proof. $\eta^k \to 0$ implies $\bar{\nu} = 0$ in (10).

To summarize, inexact minimization may worsen the rate of convergence, but if the stopping rule is operated with $\eta^k \to 0$ (sufficiently rapidly), inexact minimization retains the rate of convergence obtainable with exact minimization.

8. Geometric Interpretations

The Multiplier Method can be explained in terms of some very instructive geometric interpretations. With the aid of two or three figures, the reader can very quickly grasp the underlying mechanism of the algorithm. In fact, one can deduce many of the algorithm's convergence properties using informal geometric proofs. At the same time one gains useful insight into the notion of ordinary Lagrange multipliers and generalized Lagrange multipliers. Finally one may perceive the multiplier method to be a hybrid between ordinary primal dual methods (based on the ordinary Lagrangian) and ordinary exterior penalty methods. In this connection, the geometric interpretation clearly shows the superiority of the multiplier method over the usual pure penalty approach.

The Primal Functional

Also called the perturbation or optimal response function, the primal functional arises by embedding the CPP (1) into a family of problems parameterized by a perturbation vector $u \varepsilon R^d$:

$$q(u) = \inf f_0(x)$$

$$\text{subject to} \quad f_i(x) \leq u_i \leq \quad i = 1,\ldots,m$$

$$f_i(x) = u_i \quad i = m+1,\ldots,d.$$

The primal functional q is convex. The original problem corresponds to zero perturbation ($u = 0$). The Lagrange multipliers are associated with hyperplanes which support q at $u = 0$. (If H^* is such a support hyperplane in Figure 1, then $-y^*$ is the gradient of the affine function whose graph is H^*.)

The next proposition shows how minimization of either Lagrangian can be interpreted in terms of a corresponding support problem on the primal functional.

Proposition 6. Let $y \varepsilon Y$. Then

$$\inf_x L(x;y) = \inf_u \{q(u) + \langle u,y \rangle\} \tag{11}$$

$$\inf_x L_r(x;y) = \inf_u \{q(u) + h_r[u;y]\} \ . \tag{12}$$

where

$$h_r[u;y] = \sum_{i=1}^{m} rp[u_i/r;y_i] + \sum_{i=m+1}^{d} y_i u_i + r\phi[u_i/r].$$

Furthermore if $\hat{x}$ minimizes $L(\cdot;y)$ (resp. $L_r(\cdot;y)$), then $\hat{u}$ minimizes the right side of (11) (resp. (12)), where

$$\hat{u}_i = f_i(\hat{x}) \qquad i = 1,\ldots,d.$$

Proof. See [5, Section 2 and prop. 4.1].

Minimizing the ordinary Lagrangian $L(\cdot;y)$ is equivalent to locating a point $(u,q(u))$ where the hyperplane $H = \{(u,w) \varepsilon R^{d+1} \mid w = \beta - \langle u,y \rangle\}$ supports q. The support hyperplane H intercepts the vertical axis at $\beta = g(y)$.

Minimizing the penalized Lagrangian $L_r(\cdot;y)$ can be viewed in the same way except that the hyperplane H has now been replaced by a concave hypersurface. The hypersurface is the graph of the function

$$u \mapsto \alpha - h_r[u;y]$$

where

$$\alpha = g_r(y) \quad \text{and}$$

$$h_r[u;y] = \sum_{i=1}^{m} rp[u_i/r;y_i] + \sum_{i=m+1}^{d} y_i u_i + r\phi[u_i/r].$$

The function $h_r[\cdot;y]$ is sometimes called a generalized Lagrange multiplier function. Note that $\nabla_u h_r[u;y]\Big|_{u=0} = y$. Thus the generalized Lagrange multiplier function $h_r[\cdot;y]$ closely approximates the ordinary (linear) Lagrange multiplier function $\langle \cdot,y\rangle$ near $u = 0$.

Figure 2 shows the supporting hypersurface interpretation for the single constraint case m=d=1. At the kth unconstrained minimization one has $y = y^k$, an estimate of the Lagrange multiplier. The hypersurface supports the primal functional q at a point u^k other than zero. The intercept $\alpha = g_r(y^k)$ lies below the optimal value $q(0)$.

At the point of support $(u^k,q(u^k))$ one can insert a hyperplane H which separates the two graphs. The hyperplane is just the tangent plane to $\alpha-h_r[\cdot;y^k]$ at $u = u^k$. Thus the separating hyperplane corresponds to an ordinary Lagrange multiplier

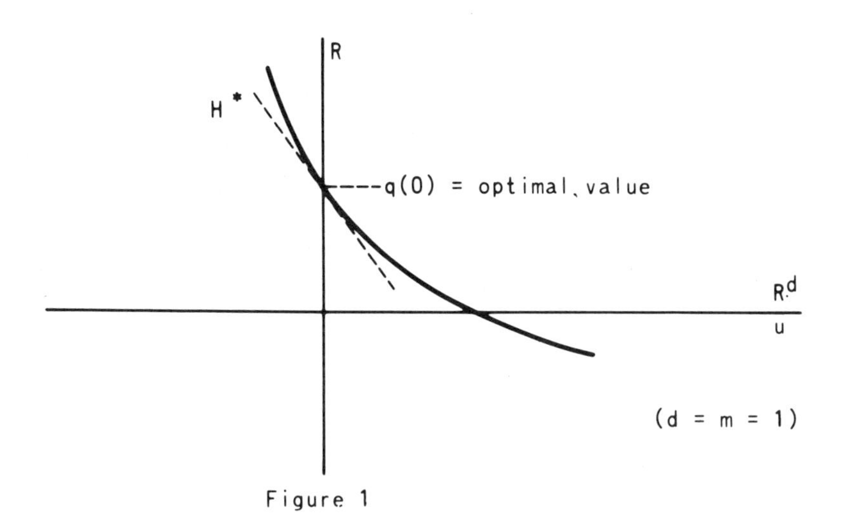

Figure 1

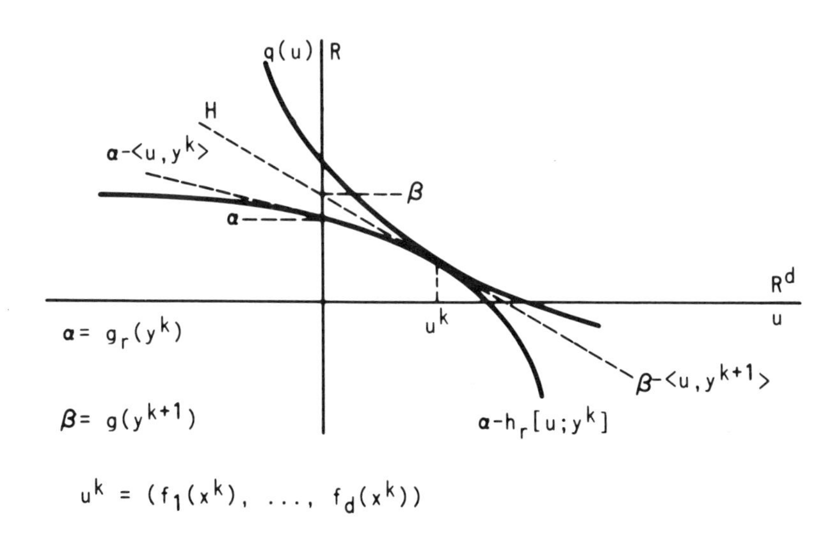

Figure 2

$\tilde{y} = \nabla_u h_r[u^k;y^k] = y^{k+1}$. The separating hyperplane intercepts the vertical axis at $\beta = g(y^{k+1})$. Note that $\beta > \alpha$; i.e., $g(y^{k+1}) > g_r(y^k)$.

The dual iteration simply replaces the function $h_r[\cdot;y^k]$ by $h_r[\cdot;y^{k+1}]$. That is, the hypersurface is adjusted so that its nominal gradient at $u = 0$ is $-y^{k+1}$. The new hypersurface will support q with an intercept $\alpha = g_r(y^{k+1})$ which lies above the intercept $\beta = g(y^{k+1})$ for the hyperplane of the previous stage. Thus the intercepts α and β climb the vertical axis in leapfrog fashion, converging to the optimal value q(0). At the same time the dual variables y^k converge to a Lagrange multiplier so that the limiting hyperplanes and hypersurfaces support q at $u = 0$.

<u>The Dual Functional</u>

Corresponding to proposition 6 is the associated conjugate expression which is based on Fenchel's Duality Theorem.

<u>Proposition 7</u>. Let $y \varepsilon Y$; then

$$g_r(y) = \max_{s \varepsilon R^d} \{g(s) - h_r^*[s;y]\} \tag{13}$$

where

$$h_r^*[s;y] = \max_u \{\langle u,s \rangle - h_r[u;y]\} .$$

<u>Proof</u>. See [5, prop. 4.2].

The function $h_r^*[\cdot;y]$ is the convex conjugate of $h_r[\cdot;y]$. There is a one-one correspondence between closed convex functions and their conjugates. The conjugate $h_r^*[\cdot;y]$ is graphed in Figure 3.

The maximization of (13) can be illustrated using the same supporting hypersurface point of view utilized in the primal space interpretations. We seek a point $s \varepsilon R^d$ and a scalar α such that the graph of $s \mapsto \alpha + h_r^*[s;y]$ supports the graph of $g(s)$ at $s = \tilde{s}$. Then $\tilde{s}$ is the maximizing point in (13) and $\alpha = g_r(y)$.

Figure 4 shows the solution to the support problem associated with (13). With $y = y^k$, the (unique) point of support is $\tilde{s} = y^{k+1}$. The conjugate function h_r^* may be thought of as a "probe." It is centered over the dual functional g at a point y^k which is an estimate of the maximizing vector of g. The probe is lowered until it contacts g. The point of contact is unique because $h_r^*[\cdot;y^k]$ is strictly convex. The point of contact is y^{k+1} which is a better estimate of the maximizing point of g. Thus the dual iteration recenters the probe over y^{k+1} and the process repeats. The points of contact climb the dual functional and converge to a Lagrange multiplier (i.e., a maximizing vector for g).

The conjugate h_r^* is directly proportional to r, the penalty parameter. As $r \to 0$ the probe becomes flatter (going to a horizontal hyperplane

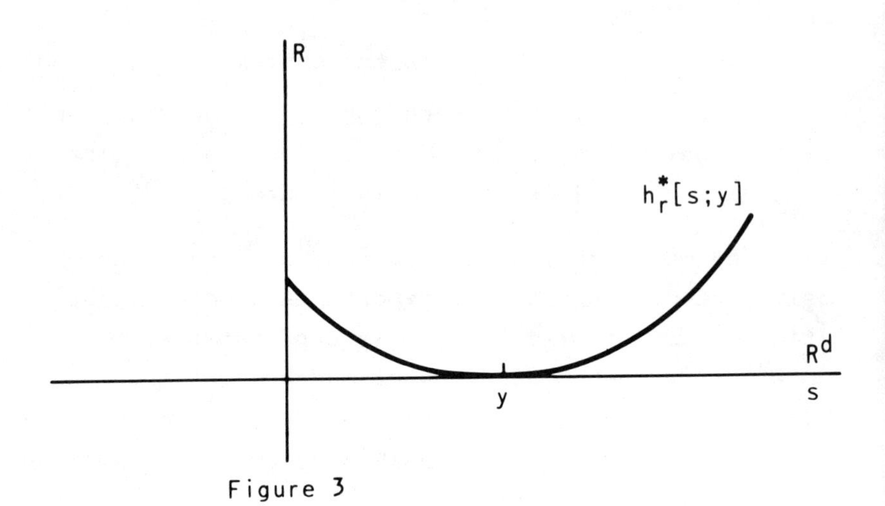

Figure 3

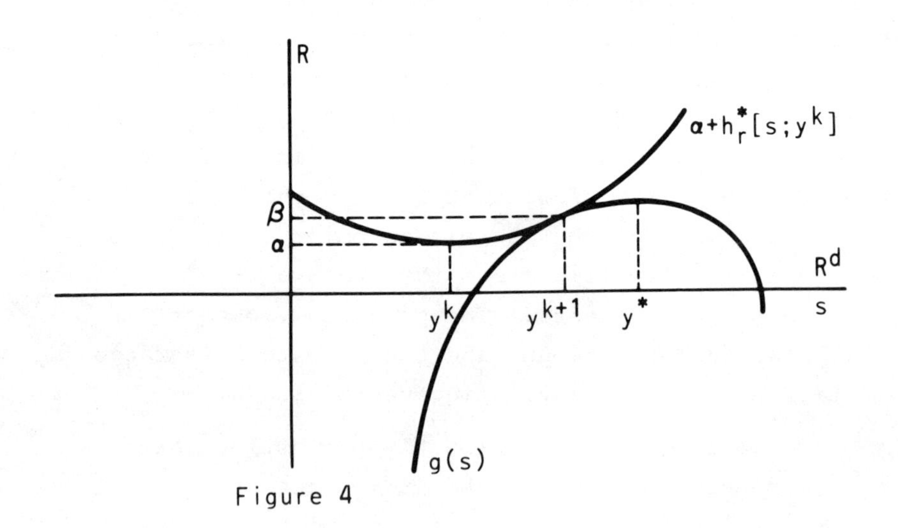

Figure 4

in the limit). Thus convergence is speeded as r is made small. Note however that small values of r worsen the condition number of the penalized Lagrangian so that the unconstrained minimizations of $L_r(\cdot;y)$ become more difficult. Conversely as r is made large, the probe becomes narrow and pointed, collapsing to a "needle" or delta-function as $r\to\infty$. The limiting case $r=\infty$ corresponds to minimizing the ordinary Lagrangian to obtain the ordinary dual g(y).

Ordinary penalty methods may be compared to the multiplier method by observing that the penalty method consists of sequentially minimizing $L_r(x;0)$ for ever smaller values of r. (The dual variable $y \equiv 0$ in the penalty approach and there is no dual iteration.) In the context of Figure 4, one keeps the probe always centered at s = 0. In successive iterations the probe is made ever flatter by reducing r. It is clear that the penalty approach cannot do as well as the multiplier method which permits recentering the probe at the most recent point of contact. This point has been substantiated analytically in [5, Section 8].

9. Conclusion

The multiplier method has been defined for a large class of penalty and barrier functions. The method is applicable to problems with both equality and inequality constraints. For convex problems the method is globally convergent. The rate of convergence is seen to depend on the choice of penalty function. The rate is linear for the "usual" penalties but higher orders of convergence are possible. The method can be operated with inexact calculations without destroying the convergence properties if a certain stopping criterion is adhered to. The method has been explained via geometric interpretations in both the primal and dual spaces. By these interpretations, one can view the method as a hybrid between ordinary primal-dual methods and ordinary exterior penalty methods.

References

[1] Arrow, K. J. and Solow, R. M. (1958) "Gradient Methods for Constrained Maxima, with Weakened Assumptions," Chapter 11 in K. J. Arrow, L. Hurwicz, and H. Uzawa, Studies in Linear and Nonlinear Programming, Stanford University Press, pp. 166-176.

[2] Hestenes, M. R., (1969) "Multiplier and Gradient Methods," Journal of Optimization Theory and Applications, Vol. 4, No. 5, pp. 303-320.

[3] Kort, B. W. and Bertsekas, D. P. (December 1973) "Multiplier Methods for Convex Programming," Proceedings of 1973 IEEE Conference on Decision and Control, San Diego, California, pp. 428-432.

[4] Kort, B. W., and Bertsekas, D. P. (August, 1973) "Combined Primal Dual and Penalty Methods for Convex Programming," Working Paper, submitted to SIAM Journal on Control.

[5] Kort, B. W. (to appear) "Combined Primal-Dual and Penalty Function Algorithms for Nonlinear Programming," Ph.D. Thesis.

[6] Powell, M. J. D. (1969) "A Method for Nonlinear Constraints in Minimization Problems," Chap. 19 in: R. Fletcher, (Ed.), Optimization, Academic Press, pp. 283-298.

[7] Rockafellar, R. T. (1973) "The Multiplier Method of Hestenes and Powell Applied to Convex Programming," Journal of Optimization Theory and Applications, Vol. 12, No. 6, pp. 555-562.

OPTIMIZATION WITH CORNERS

by

A. A. Goldstein

The discussion presented here was motivated by a "Penalty" method of Rockafellar. Roughly, we are given a function with corners and we are seeking a stationary point. The stationary point may be either a corner point or a smooth point; an algorithm is desired which will work in either case. We discuss simple problems of this type.

We shall dwell on the approach by "descent", however, two other approaches will be considered briefly. A general theory of descent first appeared in [2], 1966. See also [3]. Since then, there have been numerous refinements (see [4] for bibliograhy to 1971) but the assumption of a uniformly continuous differential has not been relaxed. In the treatment below we relax to one-sided differentials so that corners can be handled for functions of polyhedral type. Problems of this type were first discussed by Demjanov in [1].

In what follows let E denote a normed-linear space (usually E_n) and let f be a real-valued function defined on E. We denote by $f'(x,h)$

It is a pleasure to thank the referees for their kind help. This study was supported by the National Science Foundation under Grant No. MPS72-04787 A02.

the Gateaux differential of f at x in the direction h. We shall assume that if f' exists that it is a bounded linear operator. Denote by $f'_+(x,h)$ the right handed Gateaux differential. This exists if the limit

$$\lim_{t\to 0+} (f(x+th) - f(x))/t$$

exists for all $h \in E$. The differential $f'_+(x,h)$ is positively homogeneous in h. If $\inf_{||h||=1} f'_+(x,h) = \mu(x)$ is achieved for some $\hat{h} \in E$, then $\hat{h}$ is called a direction of steepest descent. With the right hand differential, stationary points can be defined to encompass corners or smooth points. A point z is called a <u>stationary point</u> if $f'_+(z,h) \geq 0$ for all $h \in E$, and strict inequality holds for some $h \in E$. <u>A partial stationary point</u> is a point z at which either $f'(z,\cdot) = 0$ or f' is discontinuous at z.

Since the direction of steepest descent may not be defined, or because other directions may be more favorable, we define a direction mapping ϕ with the property that $f(x+t\phi(x)) < f(x)$ for small t.

Take x_o arbitrarily in E. Let $S = \{x: f(x) \leq f(x_o)\}$. Let $S' = \{x \in S: x$ is stationary$\}$. Let ϕ denote a (possibly) set-valued bounded mapping from S to the set of subsets of E which satisfies:

1) $f'_+(x,\phi(x)) < 0$ for $x \in S \sim S'$

2) If $\inf_{h\in\phi(x_k)} f'_+(x_k,h) \to 0$ then $\mu(x_k) \to 0$.

The second condition is imposed for the smooth case. For corners, $f'_+(x_k, \phi(x_k))$ will be bounded away from 0.

Example: Assume $f'_+(x,h)$ is continuous in h for $h \neq 0$. Choose $\theta(x)$, $\|\theta(x)\| = 1$ such that $f'_+(x,\theta(x)) \leq \alpha\mu(x)$, where $0 < \alpha \leq 1$. Set $\phi(x) = \theta(x)|\mu(x)|$ then $f'_+(x,\phi(x)) \leq \alpha\mu(x)|\mu(x)|$ and (2) above holds.

As a model problem we consider the following generalization of a convex polyhedral function.

Let $I = \{1,2,\ldots,m\}$. Let a^i, $i \in I$ be real-valued functions defined on E_n and set $f(x) = \max\{a^i(x): i \in I\}$. Assume that on S a^i has lip continuous gradients. A facet of f is the set $F = \{x,f(x)): x \in S,\ f(x) = a^{i}0(x)$, and $a^i(x) < f(x),\ i \in I \sim i_0\}$. If $(x,f(x)) \in F$ and $\nabla f(x) = 0$ we call x a critical point. We assume the number of critical points is finite. If $x \in S$ is not a critical point, we assume that every subset of n points of $\{\nabla a^i(x): i \in I\}$ and $n + 1$ points of $\{(\nabla a^i(x),1): i \in I\}$ respectively, are linearly independent. Let $I(x)$ denote that subset of I for which $f(x) = a^i(x)$. A vertex of f is a point $(x,f(x))$ where x satisfies:

$$a^i(x) = f(x) \quad i \in I(x),\ \text{card } I(x) \geq n + 1$$
$$a^i(x) < f(x) \quad i \in I \sim I(x)$$

A vertex may be a partial-stationary or stationary point. We assume vertices exist, their total number is finite, and that any stationary point of f is a critical point or vertex. Any f satisfying these properties will be called a quasi simplicial polyhedral function. If the a^i are affine functions f is a simplicial convex polyhedral function. Let $H(S)$ denote the convex hull of S.

Theorem: There exists a sequence of positive numbers $\{\gamma_k\}$ and a mapping ϕ such that if

$$x_{k+1} = x_k + \gamma_k \phi(x_k) \quad \text{and}$$

1) if $f'_+(x,\cdot)$ exists for $x \in H(S)$ and $f'_+(x,h)$ is continuous in h for $h \neq 0$, if S and $\overline{\phi(S)}$ are compact, if $f'_+(x,h) \le M\|h\|$ for all $x \in S$, and if z and $\hat{\phi}$ are cluster points of $\{x_k\}$ and $\phi(x_k)$ respectively, then z is a partial stationary point.

2) If f is a simplicial convex polyhedral function on E_n which is bounded below, the sequence $\{x_k\}$ converges to a minimum of f in a finite number of steps.

3) If f is a quasi simplicial polyhedral function on E_n such that S is bounded, then $\{x_k\}$ converges to a vertex or critical point of f.

Proof:

(1) Choose ϕ as in the example above. Choose γ_k to minimize $f(x_k + \gamma\phi(x_k))$. Take subsequences $\{x_k\}$ and $\{\phi_k\}$ converging to z and $\hat{\phi}$ respectively. If $f'_+(z,\hat{\phi}) < 0$, choose γ_0 to minimize $f(z+\gamma\hat{\phi})$.

Then $f(z) = f(z+\gamma_o\hat{\phi}) + 2\varepsilon$ where $\varepsilon > 0$, . Set $a = z + \gamma_o\hat{\phi}$ $b_k = x_k + \gamma_o\phi(x_k)$. By a mean value theorem of McLeod [5], $(f(b_k) - f(a))/ \|b_k-a\|$ belongs to the convex closure of the set $\{f'_+(x,(b_k-a)/ \|b_k-a\|): \ x \in L(a,b_k)\} = C^k$, where $L(a,b_k)$ denotes the line segment between a and b_k . Since $z + \gamma_o\hat{\phi}$ belongs to the interior of S $L(a\ b_k)$ will frequently be contained in $H(S)$. For such k take a point in the convex hull of C^k within 1 of $(f(b_k) - f(a)/ \|b_k-a\|$. By Karatheodori's theorem, represent this point as the convex combination $\lambda_1^k f'_+(\xi_1^k,\ (b_k - a)/ \|b_k-a\|) +$ $\lambda_2^k f'_+(\xi_2^k,(b_k-a)/ \|b_k-a\|)$ where $\xi_i^k \in L(z+\gamma_o\hat{\phi},$ $x_k+\gamma_o\phi(x_k)$ belong to $H(S)$. Thus $f(b_k) \le f(a) + \|b_k - a\| (M+1)$ or $f(x_k + \gamma_o\phi(x_k) \le$ $f(z+\gamma_o\hat{\phi}) + (M+1)(\|x_k-Z\| -\gamma_o \|\hat{\phi}-\phi(x_k)\|)$. Take k so large that this last term is less than ε . Then $f(z) \le f(x_k+\gamma_k\phi(x_k)) \le f(x_k+\gamma_o\phi(x_k) \le$ $f(z+\gamma_o\hat{\phi}) + \varepsilon = f(z) - \varepsilon$. This contradiction shows that $f'_+(z,\hat{\phi}) \ge 0$.

Assume now that f' is continuous at z and $f'(z,\cdot) \ne 0$. Let N be a neighborhood of z where f' is defined. Take sequences $\{x_k\}$ and $\{\phi(x_k)\}$ such that $x_k \in N$ and $\{x_k\}$ and $\{\phi(x_k)\}$ converge to z and $\hat{\phi}$ respectively. Then

$$f'(z,\hat{\phi}) - f'(x_k,\phi(x_k)) = f'(z,\hat{\phi}) - f'(x_k,\hat{\phi}) -$$

$$f'(x_k,\phi(x_k)) + f'(x_k,\hat{\phi}) \le f'(z,\hat{\phi}) - f'(x_k,\hat{\phi})$$

$$+ M \|\hat{\phi} - \phi(x_k)\| .$$

Hence $\{f'(z,\hat{\phi}) - f'(x_k,\phi(x_k))\} \to 0$, contradicting the inequalities $f'(z,\hat{\phi}) > 0$ and $f'(x_k,\phi(x_k)) < 0$. Therefore z is stationary.

(2) Consider the problem of minimizing a simplical convex polyhedral function. Let $I = \{1,2,3,\ldots,m\}$. Let A^i and x denote pointe in E_n , b_i real numbers, and $[\ ,\ ]$ the inner product in E_n . Set

$$f(x) = \max_{i\varepsilon I} \{[A^i,x] - b_i\} = \max_{i\varepsilon I} a^i(x) .$$

Given a point x_k take $I(x_k) \subset I$ such that

$$\begin{aligned} f(x_k) &= a^i(x_k) \qquad i \ \varepsilon\ I(x_k) \\ &> a^i(x_k) \qquad i \ \varepsilon\ I \sim I(x_k) \end{aligned}$$

Let card $(I(x_k)) = p \le m$. If x_k is not a stationary point of f then the system of inequalities

$$[A^i,\phi] \le 1 \qquad i \ \varepsilon\ I(x_k) \tag{1}$$

is consistent. Let $I'(x_k) \subset I(x_k)$ be such that $\{A^i\colon i \ \varepsilon\ I'(x_k)\}$ is a maximal subset of linearly independent points of $\{A^i\colon i \ \varepsilon\ I(x_k)\}$. Thus, if $p \le n$, $I'(x_k) = I(x_k)$.

By a theorem of Fan, [6], there exists $\phi(x_k)$ such that

$$[A^i, \phi(x_k)] = -1 \quad i \in I'(x_k)$$
$$[A^i, \phi(x_k)] < -1 \quad i \in I(x_k) \sim I'(x_k)$$

As in 1) above, choose γ_k to minimize $f(x_k + \gamma\phi(x_k))$. Consider an application of one step of the algorithm, going from x_k to x_{k+1}. We have

$$a^i(x_k) = f(x_k) \quad i \in I(x_k)$$
$$a^i(x_k) < f(x_k) \quad i \in I \sim I(x_k) \ .$$

Suppose card $(I(x_k)) = p$. On the ray $x = x_k + \gamma\phi(x_k)r$ p functions $a^i(x)$, $i \in I'(x_k)$, are decreasing at the same rate while the remaining functions $a^i(x)$, $i \in I(x_k) \sim I'(x_k)$ are decreasing faster. At γ_k at least one affine function $a^{i_1}(x)$, $i_1 \in I \sim I(x_k)$ which increases with γ becomes equal to those which are decreasing equally. Then card $I(x_{k+1}) \geq r + 1$. After n steps card $I(x_k) \geq n + 1$ and $(x_k, f(x_k))$ is a vertex of f. The number of vertices being finite the conclusions follow:

(3) We observe first that $f'_+(x,h)$ exists and is continuous in h for $h \neq 0$. Since $\frac{f(x+th) - f(x)}{t} = \frac{1}{t}[\max\{a^i(x+th):$ $i \in I\} - \max\{a^i(x): i \in I\}] = \frac{1}{t}\{a^{i_1}(x+th) - a^{i_o}(x)\}$,

we have for t sufficiently small that $i_1 \varepsilon I(x)$. Hence $\frac{f(x+th) - f(x)}{t} = \frac{1}{t} \max\{a^i(x+th) - a^i(x):$ $i \varepsilon I(x)\}$. Interchanging max and lim, we get $f'_+(x,h) = \max\{[\nabla a^i, h]: i \varepsilon I(x)\}$.

We consider next the definition of ϕ. There will be 2 cases:

Case 1. $\phi(x) = -\nabla f(x)$

Case 2. $\phi(x)$ satisfies the system of equations $[\nabla a^{i_j}(x), \phi(x)] = -1,\ 1 \leq j \leq n$,

where
$a^{i_1}(x) = f(x) \geq a^{i_2}(x) \geq, \ldots, \geq$
$a^{i_n}(x) \geq a^{i_j}(x),\ j \geq n+1$.

Corresponding to $\phi(x)$ a number γ is chosen as follows. Set $g(x,\gamma) = \frac{f(x+\gamma\phi(x))-f(x)}{\gamma f'_+(x,\phi(x))}$, and choose σ, $0 < \sigma < \frac{1}{2}$. Since $\lim g(x,\gamma) = 1$, $g(x,\gamma) > 0$, for small positive γ. Thus for this γ, $x + \gamma\phi(x) \varepsilon S$ and $g(x,\cdot)$ is continuous and remains continuous while it is positive. If $g(x,1) < \sigma$, choose γ so that $\sigma \leq g(x,\gamma) \leq 1 - \sigma$, this is possible because of the continuity of $g(x,\cdot)$. If $g(x,1) \geq \sigma$, set $\gamma = 1$.

Assume x_k and $\phi(x_k)$ are given. For a trial, set $\phi(x_k) = \nabla f(x_k)$. If for any ξ on the line segment joining x_k and x_{k+1} $(\xi, f(\xi)) \varepsilon F$, a facet of f, $\phi(x_k)$ is chosen from Case 1, otherwise by Case 2. We have then that

$$f(x_k) - f(x_{k+1}) = -\gamma_k f'_+(x_k, \phi(x_k)) g(x_k, \gamma_k) \geq$$

$$-\gamma_k f'_+(x_k, \phi(x_k)) \sigma.$$

If $\gamma_k f'_+(x_k, \phi(x_k)) \nrightarrow 0$, then $f(x_k) \downarrow -\infty$, contradicting that f is bounded below.

Suppose that Case 1 arises infinitely often. Let $\{x_k\}$ be a subsequence such that $\phi(x_k)$ is always chosen from Case 1. Let z be cluster point of $\{x_k\}$. Since ∇f is defined on the line segment joining x_k and x_{k+1} then

$$g(x_k, \gamma_k) = \frac{f(x_{k+1}) - f(x_k)}{\gamma_k [\nabla f(x_k), \phi(x_k)]} =$$

$$1 + \frac{[\nabla f(\xi_k) - \nabla f(x_k), \nabla f(x_k)]}{[\nabla f(x_k), \phi(x_k)]}$$

where ξ_k is between x_k and x_{k+1}. Since ∇f is lip continuous and bounded on each facet, if $\{\gamma_k\} \to 0$ then $g(x_k, \gamma_k) \to 1$, a contradiction. We are led therefore to the assumption that $\{\gamma_k\}$ is bounded away from 0. But if $\{\gamma_k\}$ is bounded away from 0 we contradict that $\{\gamma_k f'_+(x_k, \phi(x_k))\} \to 0$, unless $\{\nabla f(x_k)\} \to 0$. Thus $\{x_{k+1} - x_k\} \to 0$. If Case 2 never arises then for the original sequence $\{x_k\}$, $\{x_{k+1} - x_k\} \to 0$. The critical points being finite in number, $\{x_k\} \to z$.

If Case 2 happens infinitely often, then for a subsequence $\{x_k\}, f'_+(x_k, \phi(x_k)) = -1$, $\{\gamma_k\} \to 0$, and therefore $\{x_{k+1} - x_k\} \to 0$. Hence for the original sequence $\{x_{k+1} - x_k\} \to 0$. Now take a subsequence such that $\phi(x_k)$ is always taken from Case 2. Let z be a cluster point. If z is a critical point we are done, by the above argument. Assume that z is not a critical point and consider the totality of solutions of

$$[\nabla a^i(x_k), \theta] = -1 \qquad i \in J \tag{A}$$

where J is any subset of n points of I . Each system (a) is solvable, and for each J , $\|\theta\|$ is a continuous function of x_k . Since $\{x_k\} \cup \{z\}$ is compact $\|\theta\|$ achieves a maximum for each J , and the totality of different sets J is finite.

Thus, the points $\phi(x_k)$ are uniformly bounded, say by M . Let $N = \max\{\|\nabla a^i(x)\| : x \in S$ and $i \in I\}$, and $\|\nabla a^i(x) - \nabla a^i(y)\| \le K\|x - y\|$ for all $i \in I$.

We now show that $f'(x, \phi(x_k))$ eventually changes sign between x_k and x_{k+1} .

Suppose that $f'(x, \phi(x_k))$ remained negative on the ray joining x_k and x_{k+1} infinitely often in k . Using Mc Leod's mean value theorem we see that f is lip-continuous. Thus f may be written as the integral of its derivative and

$$g(x,\gamma) = \int_0^\gamma \frac{f'_+(x+\theta\phi(x), \phi(x))}{\gamma f'_+(x, \phi(x))} \, d\theta \ . \quad \text{Since}$$

$f'_+(x+\theta\phi)(x), \phi(x)) < 0$, the integrand is positive.

Since $\lim_{\gamma \to 0+} g(x,\gamma) = 1$, $g(x,1) > 1$ and the choice $\gamma_k = 1$ will be made. This contradiction shows that for k sufficiently large $f'_+(x,\phi(x_k))$ will change sign between x_k and x_{k+1}.

Given $\varepsilon > 0$ choose k so large that $\|x_{k+1} - x_k\| < \min(\varepsilon/2N, 1/KM)$ and $f'(x,\phi(x_k))$ changes sign between x_k and x_{k+1}. Consider now a cycle of the algorithm. We have for $1 \le j \le n$ and some ξ_k between x_k and x_{k+1} that $a^{i_j}(x_{k+1}) - a^{i_j}(x_k) = -\gamma_k + \nabla_k[\nabla a^{i_j}(\xi_k) - \nabla a^{i_j}(x_k), \phi(x_k)] \le \gamma_k[-1+K\|x_{k+1} - x_k\| M] < 0$. On the other hand since $f'_+(x,\phi(x_k))$ changes sign on the ray joining x_k and x_{k+1} $f(x_k + \gamma\phi(x_k)) = a^{i_r}(x_k + \gamma\phi(x_k))$ for some γ and $r \ge n+1$. Thus $a^{i_r}(x_{k+1}) \ge a^{i_1}(x_{k+1})$. Now $a^{i_r}(x_{k+1}) - a^{i_r}(x_k) \le N\|x_{k+1} - x_k\|$ so that $a^{i_r}(x_{k+1}) - \min_{1 \le j \le n} a^{i_j}(x_{k+1}) \le 2N\|x_{k+1} - x_k\| < \varepsilon$. Thus z is a vertex. Since the number of vertices and zeros of ∇F are finite and $\{x_{k+1} - x_k\} \to 0$, the original sequence $\{x_k\}$ must converge to z.

Approach I, (Bypassing a vertex)

Detour Theorem. Let a sequence $\{x_k\}$ be defined as in 3 of the above theorem. Assume $\{x_k\} \to z$ and that z is a vertex which is not stationary. Assume card $I(z) = n + 1$. A finite sequence $\{y_k: k = 1,2,3,\ldots,s\}$ can be constructed such that $y_k = x_k$, $k \leq s - 1$ and $f(y_s) < f(z)$.

Proof. At z, $a^i(z) = f(z)$ for all $i \in I(z)$, and z satisfies the equation

$$a^i(x) - M = 0 \qquad \text{(I)}$$

where $(x,M) = (z,f(z))$. For each k we are given an approximate solution to this system, namely $(x_k,f(x_k))$. We now estimate $f(x_k) - f(z)$. The equation (I) may be written as the operator equation

$$F(y) = 0$$

where $y = (x,M)$ and $F(y) = a^i(x) - M \quad i \in I(z)$. The range and domain of F are subsets of E_{n+1}. The operator $F'(y,\cdot)$ can be represented by the matrix whose rows are $(\nabla a^i(y), -\delta^i)$ with $\delta^i = 1$ and $i \in I(z)$. Thus the inverse of $F'(y,\cdot)$ exists and is continuous provided y is not a critical point. Let $F'_{-1}(y,\cdot)$ denote this inverse. Let ℓp norms in E_{n+1} be denoted by $\|\cdot\| p$. The subscript 2 will be suppressed. Because of the equivalence of these norms on E_{n+1} a lip condition for a function in one p implies a lip condition holds for the function for all p.

We calculate:

$||F'(x,\cdot) - F'(y,\cdot)||_\infty = \max_i ||\nabla a^i(x) - \nabla a^i(y)||_1 \le$ $\max_i ||\nabla a^i(x) - \nabla a^i(y)||_\infty \le C\,||x - y||_\infty$. Thus for some constant K :

$$||F'(x,\cdot) - F'(y,\cdot)|| \le \frac{K}{2}\,||x - y|| \ .$$

By Kantorovich's theorem [3] p. 14], if $n_k = ||F'_{-1}(y_k)F(y_k)||$ and $\beta_k = ||F'_{-1}(y_k)||$ then $\beta_k K n_k \le \frac{1}{2}$ implies that F has a root in the ball:

$$B = \{y: \ ||y - y_k|| \le 2n_k\} \ .$$

Whence $f(x_k) - f(z) = |f(x_k) - f(z)| \le 2n_k$.

Since z is not a critical point we may assume that $F'_{-1}(y_k)$ is uniformly bounded. Because $F(y_k) \to 0$ we have $\{\beta_k K n_k\} \to 0$. Suppose now that $R^{i_0}(z) = \max\{a^i(z): i \in I \sim I(z)\}$. Set $f(z) - R^{i_0}(z) = \varepsilon$. Consider the solution z_k to the system $[\nabla a^i(x_k), z_k] \le -1,\ i \in I(z)$. Let θ bound $||z_k||$ for all k . Choose $r \ge 2$ so that $\gamma = \frac{1}{rK\theta^2} < \frac{\varepsilon}{N\theta}$. Choose k so large that $\beta_k K n_k \le \frac{1}{2}$ and $2n_k < \frac{r-1}{r}\,(\frac{1}{rK\theta^2}) = \alpha$. Set $z_k = \phi$. Then

$a^{i_0}(x + \gamma\phi) = f(z) - \varepsilon + \gamma[\nabla a^{i_0}(\xi),\phi] \le$ $f(z) - \varepsilon + \gamma N\theta < f(z)$. Also $a^i(x_k) - a^i(x_k + \gamma\phi) \ge$ $\gamma[1 - K\gamma\phi^2] = \alpha \quad i \in I(z)$. Thus

$a^i(x_k + \gamma\phi) < a^i(x_k) - \alpha \le f(x_k) - \alpha < f(z)$.

Remark: Acceleration can be applied in both cases of 3 in the theorem above, as in [3] p. 153.

Approach II

While the ideas below are applicable to our model problem, we consider for simplicity the related problem of minimizing $f(x) = \max_i |a^i(x)|$.

Suppose that if $(z,f(z))$ is a vertex of f , and card $I(z) = n + 1$. Approximate f by the function $f_p^{1/p}(x) = [\sum_{r=1}^{m} a^i(x)^p]^{1/p}$ where p is a large even integer. Then apply descent to the differentiable function f_p .

Let us consider the case when f_p has only one stationary point x_p . Then $(x_p f_p(x_p))$ is a lowest point of f_p . Let $I(x_p)$ denote the indices of the largest $n + 1$ numbers $a^i(x_p)$. If the system

$$\operatorname{sgn}(a^i(x_p))[\nabla a^i(x_p),\phi] \leq -1 \qquad (I)$$

is inconsistent, then $I(x_p) = I(z)$ where z minimizes f . If (I) is consistent replace p by $2p$ and find the lowest point of f_p again. Repeat until (I) is inconsistent. The lowest point z can now be obtained by applying 3 of the above Theorem, with no "detour" necessary. Alternatively, x is an approximation of z , and distance to z can be checked by the Kantorovich Theorem as above.

This method has been successfully applied to the case when a^i is affine [7].

Approach III, The Convex Case

Assume f convex and S bounded.

In this case we can with the aid of descent formulate an algorithm which gives at each cycle upper and lower bounds for the minimum. These bounds decrease and increase respectively at each cycle. Steepest descent will be combined with an algorithm of [8]. The terminology follows the exposition in [3].

Assume that at each point $(\xi, f(\xi))$ a supporting hyperplane for f is given: $\{(x,z): x \in E_n$ and $z = [A(\xi),x] - b(\xi)]\}$, and that a set of points $x_1, x_2, \ldots, x_r$ is given with the property that the set $S^o(a) = \{x: [A(x_i),x] - b(x_i) \leq \alpha, 1 \leq i \leq r\}$ is bounded an non-empty for some α.

Algorithm. Assume that a set of $r - 1$ points $X^r = \{x_i \in E_n: 1 \leq i \leq r - 1\}$ is given. Let $S^r(\alpha) = \{y \in E_n: [A(x_i),y] - b(x_i) \leq \alpha, x_i \in X^r$. Assume that $S^r(\alpha)$ is bounded for some α.

At the m'th step $(m > r)$, assume X^m and Y_m given. Select x^m to minimize the function

$$f^m(x) = \max\{[A(x_i),x] - b(x_i): x_i \in X^m\}.$$

Choose $\phi(y_m)$ and γ_m as in the Theorem above. Set $y_{m+1} = y_m + \gamma_m \phi(y_m))$ if $f(y_m + \gamma_m \phi(x_m)) <$ $f(x_m)$. Otherwise set $y_{m+1} = x_m$.

Claim. $f_m(x_m) \uparrow \min f$ while $f(y_m) \downarrow \min f$. Every cluster point of $\{x_m\}$ and $\{y_m\}$ minimizes f . If f has a unique minimum these sequences converge.

Proof. The proof follows [8] with slight modifications.

Bibliography

[1] Demjanov, V. F. (1968) Algorithms for some Minimax Problems, Journal of Computer and System Sciences: 2, 342-380

[2] Goldstein, A. A. (1966) Minimizing functionals on normed linear spaces. J. SIAM Control, Vol. 4, #2, pp. 81-89.

[3] Goldstein, A. A. (1967) Constructive Real Analysis. Harper and Row, New York.

[4] Daniel, J. W. (1971) The Approximate Minimization of Functionals. Prenctice Hall, Englewood Cliffs, N. J.

[5] McLeod, R. M. (1964-65) Mean Value theorems for vector valued functions. Proc. Edinburgh Math. Soc., 14, Sec. II, pp. 197-209.

[6] Fan, Ky (1956) On Systems of linear inequalities. Linear Inequalities and related systems. Ed. Kuhn and Tucker, Princeton, pp. 99-156.

[7] Goldstein, A. A., Levine, N., and Hereshaff, J. B. (1957) On the best and least Q'th approximation of an overdetermined system of linear equations. J. Assoc. Comput. Mach. 4, pp. 371-447.

[8] Cheney, E. W. and Goldstein, A. A. (1959) Newton's method for convex programming and Tchebycheff approximation. Num. Math. 1, pp. 253-268.

THE USE OF MATRIX FACTORIZATIONS IN DERIVATIVE-FREE NONLINEAR LEAST SQUARES ALGORITHMS

by

Richard H. Bartels
The Johns Hopkins University
Baltimore, Maryland

Abstract

Algorithms based on Powell's Hybrid Method and/or the Marquardt-Levenberg scheme for minimizing the sum of squares of nonlinear functions and requiring neither analytic nor numeric derivatives are considered. These algorithms employ a pseudo-Jacobian matrix J which is updated by an unsymmetric rank-one formula at each cycle. Two factorizations, the QR factorization of J and the R^TR factorization of J^TJ, are studied, and implementational details concerned with the careful and efficient updating of these factorizations are presented. It is indicated how degeneracies may be monitored via the factorizations, and some observations are presented based upon test-code results.

1. Introduction

It frequently happens that a minimum of the sum of squares of some nonlinear functions must be found where analytic derivatives of the given functions are, for practical purposes, unobtainable. Since rapidly converging algorithms for the minimization of nonlinear sums of squares require derivatives, programs are frequently written to make use of numeric approximations to those derivatives which are otherwise unavailable. But in extreme cases the function calculations needed to produce these approximations may be too costly to be borne. The author has encountered two such situations--in the preparation of software to carry out least squares data fitting using splines with variable knots and in the determination of molecule shapes by molecular orbital calculations. Ideas presented by Powell [11] in connection with the solution of systems of nonlinear equations have been extended to handle nonlinear least squares problems with some degree of success. The resulting algorithms exhibit linear convergence, but work-per-cycle is often enough less than is required by methods using numeric derivatives that the overall cost in finding a minimum is smaller.

Some thought has been given to the efficient implementation of matrix computations required by algorithms which extend Powell's ideas in [11] to nonlinear least squares problems. This paper concentrates mainly upon implementational details of these matrix computations, and we shall remain intentionally vague about other specifics. Briefly, what is presented here has application to any cyclic algorithm which goes into its kth cycle with an $m \times n$ matrix J_k $(m \geq n)$, which occupies part of that cycle solving a linear least square problem involving J_k, and which ends the kth cycle by setting $J_{k+1} = J_k + y_k p_k^T$ for some vectors y_k (of length m) and p_k (of length n).

2. Notation and Outline of an Algorithm

Let $f(x)$ be a vector-valued function, $f(x) = [\phi_1(x), \ldots, \phi_m(x)]^T$, where each ϕ_1 is a real-valued function of $x \in \mathbb{R}^n$.

We wish to minimize

$$(2.1) \qquad f(x)^T f(x) = \|f(x)\|^2 = \sum_{i=1}^{m} \phi_i(x)^2.$$

When $m = n$ and the minimization is performed to find a zero of $f(x)$, then the Newton step is given at any point x by

$$(2.2) \qquad d = -J(x)^{-1} f(x)$$

and the steepest descent direction by

$$(2.3) \qquad g = -J(x)^T f(x),$$

where $J(x)$ is the Jacobian matrix of f; i.e., the matrix whose ith row is the gradient of ϕ_i.

Powell presents an argument for making steps from given points to new points along directions p which interpolate between d and g. And he proposes removing derivatives from the problem entirely by approximating $J(x)$ and $J(x)^{-1}$, respectively, with a quasi-Jacobian matrix J and its inverse H, both of which are updated using unsymmetric rank-one formulas. He replaces (2.2) and (2.3) by a quasi-Newton step and a quasi-steepest-descent direction, respectively, to propose an algorithm with the following outline:

(2.4) Begin with x_k, H_k, J_k, and a step size δ_k.
Set

$$d_k = -H_k f(x_k)$$

$$g_k = -J_k^T f(x_k)$$

$$p_k = \alpha_k d_k + \beta_k g_k$$ (for suitably chosen scalars α_k and β_k so that, among other things, $||p_k|| \leqq \delta_k$)

$$y_k = \frac{1}{p_k^T p_k} [f(x_k+p_k) - f(x_k) - J_k p_k]$$

$$J_{k+1} = J_k + y_k p_k^T$$

$$H_{k+1} = H_k + \frac{1}{1+p_k^T H_k y_k} H_k y_k p_k^T H_k$$

and construct x_{k+1}, δ_{k+1} in some appropriate way from x_k, δ_k.

Alternatively, one can cast out derivatives and achieve the effect of taking minimizing steps intermediate between the quasi-Newton and quasi-steepest-descent directions by using the technique of Marquardt and Levenberg, wherein the matrix H_k is not used and p_k is given instead by

$$p_k = -\gamma_k [J_k^T J_k + \lambda_k^2 I]^{-1} J_k^T f(x_k) \tag{2.5}$$

for suitably chosen parameters γ_k, λ_k (possibly determined via a linear search). Further information relating to aspects of this may be found in [1,2,10].

The formula given for J_{k+1} is such that the pseudo-Newton equation

(2.6) $$J_{k+1}p_k = f(x_k+p_k) - f(x_k)$$

is satisfied. That is, the elements of J_k are changed to make the ith element of the vector $J_{k+1}p_k$ reflect the directional derivative of ϕ_i along the vector p_k at some point (different for each i) between x_k and $x_k + p_k$. The change in J_k has been made unique by requiring that

(2.7) $$J_{k+1}h = J_k h$$

for any vector h orthogonal to p_k.

If a sequence of p vectors is produced which are restricted to some subspace of $\mathbb{R}^n$, as when the algorithm is "following a valley," this limits the chance that the J matrices will indicate the sensitivity of f to changes in the complement subspace as the actual Jacobian of f would do. Thus Powell proposes the introduction of "pseudo-iterations" in which occasional p_k are chosen to be in directions orthogonal to a number of the preceding displacements rather than being chosen according to the outline in (2.4). The use of pseudo-iterations is to be recommended strongly for any algorithm in which the pseudo-Jacobian J_k is used in place of the true Jacobian $J(x_k)$.

3. Extension to Nonlinear Least Squares

A natural extension to nonlinear least squares (m>n) is obtained by considering interpolations between the steepest descent direction g given by (2.3) and the

Gauss-Newton displacement d which satisfies

(3.1) $\quad J(x)d = -f(x)$

in the least squares sense. This would suggest that the matrix H of Powell's algorithm should be made to correspond to the generalized inverse of J in an extended version of the algorithm, and R. Fletcher [5,6] has made proposals along these lines. The corresponding updating formulas, however, have very unsatisfactory numerical properties.

Instead, we consider producing d_k to solve $J_k d_k = -f(x_k)$ in the least squares sense by using either of two matrix factorizations which facilitate the solution of (3.1). In the first we consider the factorization

(3.3) $\quad Q_k R_k = J_k,$

where R_k is an $n \times n$ right triangular matrix and the n columns of the $m \times n$ matrix Q_k form an orthonormal system of vectors; i.e., $Q_k^T Q_k$ is the $n \times n$ identity. In this case d_k is given by the computation:

$$(3.4) \quad \begin{cases} \text{set } b_k = -Q_k^T f(x_k), \text{ and} \\ \text{solve } R_k d_k = b_k \text{ for } d_k, \end{cases}$$

the justification for which may be found in substituting (3.3) into (3.1). Further details about the use of the QR factorization to solve least square problems appear in [4,9].

Alternatively, the matrix Q_k can be discarded, and J_k together with an $n \times n$ right triangular matrix R_k satisfying

$$(3.5) \qquad R_k^T R_k = J_k^T J_k$$

may be used to produce d_k from the computation:

$$(3.6) \quad \begin{cases} \text{set } c_k = -J_k^T f(x_k), \\ \text{solve } R_k^T s_k = c_k \text{ for } s_k, \text{ and} \\ \text{solve } R_k d_k = s_k \text{ for } d_k . \end{cases}$$

Again the justification for this comes from (3.1). The matrix R_k of (3.5) need not be the same as that in (3.3), but economy of notation and a close relationship between (3.1), (3.4), and (3.6) justifies our usage.

The modification of algorithm (2.4) to use either of these computations should be clear. This algorithm then becomes an appropriate base from which to attack nonlinear least squares problems.

The path either over route (3.4) or route (3.6) to d_k can be travelled at a cost of $O(n[m+n])$ arithmetic operations. The double use of R_k in (3.6), however, reflecting the explicit use of the normal equations (3.1) to solve the linear least squares problem (3.2) will mean that d_k will be produced less accurately from (3.6) than from (3.4). Great apology must therefore be given to the use of (3.5) and (3.6) by any self-respecting numerical analyst. In this regard, we offer the observation that, in our experimental codes, no minimum achieved by the use of (3.3), (3.4) to produce the Gauss-Newton displacements d_k has ever failed to be achieved by producing these displacements through the use of (3.5), (3.6). And the inaccuracies of the latter computation actually appeared to have had a beneficial effect in certain problems, yielding slightly quicker descents to the minima.

The details of Powell's algorithm [11] which select the mix between the displacement d_k and a step along the direction g_k can be carried over to an algorithm for the minimization of nonlinear sums of squares. If the Marquardt-Levenberg techniques are preferred instead, then we note with Osborne [10] that when $m > n$ the vector p_k of (2.5) satisfies the linear system

$$(3.7) \qquad \frac{1}{\gamma_k} \begin{bmatrix} J_k \\ \hline \lambda_k I \end{bmatrix} p_k = \begin{bmatrix} f(x_k) \\ \hline 0 \end{bmatrix}$$

in the least squares sense where I is the $n \times n$ identity. Either of the factorizations (3.3) or (3.5) can be used to solve (3.7), the critical step in the process being the reduction of the matrix

$$(3.8) \qquad \begin{bmatrix} R_k \\ \hline \lambda_k I \end{bmatrix}$$

to "right triangular" form by means of orthogonal transformations. This step unfortunately costs $O(n^3)$ arithmetic operations (though fewer than the $O(n^2[m+n])$ which would be demanded by the direct reduction of the matrix in (3.7) to right triangular form). Details of the computation may be found in [1].

With the above as background information, then, we shall be interested here in the mechanics of producing, in an efficient and careful way, R_{k+1} (and Q_{k+1}) in the context of algorithms whose bare outlines are:

(3.9) Given R_k ($n \times n$, right triangular) and J_k ($m \times n$) such that $R_k^T R_k = J_k^T J_k$, determine p_k (n-vector) and y_k (m-vector), set $J_{k+1} = J_k + y_k p_k^T$, and find R_{k+1} so that $R_{k+1}^T R_{k+1} = J_{k+1}^T J_{k+1}$.

(3.10) Given R_k and Q_k ($m \times n$ such that $Q_k^T Q_k = I$), determine p_k and y_k, and find Q_{k+1} and R_{k+1} so that $Q_{k+1} R_{k+1} = Q_k R_k + y_k p_k^T$.

4. Mechanics of Updating $R^T R$

If $\bar{J} = yp^T$, then

$$\bar{J}^T \bar{J} = J^T J + zp^T + pz^T + \eta pp^T \tag{4.1}$$

where

$$z = J^T y \quad \text{and} \quad \eta = y^T y. \tag{4.2}$$

We wish to derive $\bar{R}$ as accurately as possible with the information at hand so that $\bar{R}^T \bar{R} = \bar{J}^T \bar{J}$. Note that $\bar{J}^T \bar{J}$ differs from $J^T J$ by a symmetric correction of rank two. In [8] a number of techniques are given for modifying the $R^T R$ factorization under a symmetric correction of rank one, and two successive applications of any of the techniques given there could be used to obtain the matrix $\bar{R}$ which is determined by (4.1). However, we feel that corresponding single-stage processes must exist for the rank-2 case. As an example, we propose a computation below which generalizes method "C4" given in [8] and produces $\bar{R}$ at less than the cost of the two applications of C4 which the production of $\bar{R}$ would demand.

Firstly note that

$$\bar{J}^T\bar{J} = R^T(I + uv^T + vu^T + \eta vv^T)R, \tag{4.3}$$

where u and v satisfy

$$\begin{aligned} R^T u &= z \quad \text{and} \\ R^T v &= p, \end{aligned} \tag{4.4}$$

and where we expect the factor

$$I + uv^T + vu^T + \eta vv^T \tag{4.5}$$

to be positive definite.

Step 1: Construct an orthogonal matrix W so that

$$Wu = \begin{bmatrix} * \\ * \\ 0 \\ \vdots \\ 0 \end{bmatrix}, \qquad Wv = \begin{bmatrix} * \\ * \\ 0 \\ \vdots \\ 0 \end{bmatrix},$$

and $B = WR$ has zeros below the <u>second</u> subdiagonal.

Then

$$R^T W^T W(I + uv^T + vu^T + \eta vv^T)W^T WR$$

has the form

(4.7) $\quad B^TDB =$

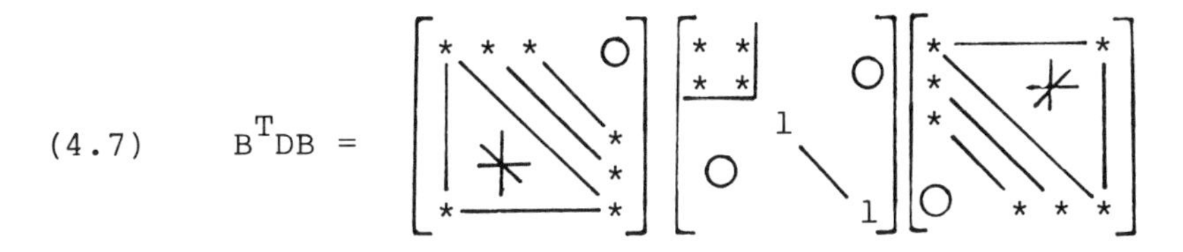

where D is equal to the identity except for its 2×2 principal minor. We expect D to be positive definite, and finding $D^{1/2}$ is an easy computation. $(D^{1/2}B)$ is a matrix with the same zero pattern as B.

<u>Step 2</u>: Construct an orthogonal matrix S so that

$$\bar{R} = S(D^{1/2}B) \tag{4.8}$$

is right triangular.

Then

$$\bar{J}^T\bar{J} = \bar{R}^T\bar{R} \ .$$

<u>Details</u>: The matrix S is constructed as a product of Householder matrices $S_{n-1}S_{n-2}\cdots S_2S_1$, chosen so that $(S_{i-1}\cdots S_1)(D^{1/2}B)$ has zeros below the diagonal in the first i-1 columns and so that the indicated subdiagonal elements:

$$S_i[(S_{i-1}\cdots S_1)(D^{1/2}B)] = S_i \begin{bmatrix} * & * & * & * & * & * \\ 0 & * & * & * & * & * \\ 0 & 0 & * & * & * & * \\ 0 & 0 & \circledast & * & * & * \\ 0 & 0 & \circledast & * & * & * \\ 0 & 0 & 0 & * & * & * \end{bmatrix} \begin{matrix} \\ \\ \\ i+1 \\ i+2 \\ \\ \end{matrix} \tag{4.9}$$

(the circled elements are in column i)

are reduced to zero by S_i.

The construction of an orthogonal matrix W satisfying all of the demands of step 1 proceeds by another sequence of simple orthogonal transformations

$W = W_3 W_4 \ldots W_{n-1} W_n$, somewhat like Householder transformations, each of which introduces a zero simultaneously into the ith positions of two given vectors:

$$(4.10) \qquad W_i \begin{bmatrix} * & * \\ * & * \\ * & * \\ \circledast & \circledast \\ 0 & 0 \\ 0 & 0 \end{bmatrix} i$$

These matrices are to be applied in succession to the vectors u and v, introducing zeros in these vectors from the bottom up.

To construct each W_i we investigate the form $I - \omega w w^T$, where $\omega = \frac{2}{w^T w}$. It is sufficient to consider the 3 × 3 case:

$$(4.11) \qquad \left(I - \omega \begin{bmatrix} w_1 \\ w_2 \\ w_3 \end{bmatrix} [w_1 w_2 w_3] \right) \begin{bmatrix} u_1 & v_1 \\ u_2 & v_2 \\ \textcircled{u_3} & \textcircled{v_3} \end{bmatrix} .$$

If w_1 and w_2 are chosen to satisfy the 2 × 2 system

$$(4.12) \qquad \begin{aligned} v_1 w_1 + v_2 w_2 &= v_3 \\ u_1 w_1 + u_2 w_2 &= u_3 \end{aligned}$$

and if w_3 is given as

$$(4.13) \qquad w_3 = -1 - \sqrt{1 + w_1^2 + w_2^2} ,$$

then $I - \omega w w^T$ will affect u and v as indicated.

This specification can fail only if (4.12) cannot be solved, which would imply that $[v_1, v_2]^T$ and

$[u_1,u_2]^T$ are co-linear. In such a case a Givens matrix can be found which will introduce a zero into the second position of these two vectors simultaneously:

$$\begin{bmatrix} c & s \\ s & -c \end{bmatrix} \begin{bmatrix} v_1 \\ v_2 \end{bmatrix} = \begin{bmatrix} * \\ 0 \end{bmatrix}$$

(4.14) and (where $c^2 + s^2 = 1$).

$$\begin{bmatrix} c & s \\ s & -c \end{bmatrix} \begin{bmatrix} u_1 \\ u_2 \end{bmatrix} = \begin{bmatrix} * \\ 0 \end{bmatrix}$$

We may then use the orthogonal matrix

$$(4.15) \qquad \begin{bmatrix} 0 & 0 & 1 \\ c & s & 0 \\ s & -c & 0 \end{bmatrix}$$

instead of $I - \omega ww^T$. for the desired effect on u and v .

In any event the matrix W will have been constructed as a product $W_3W_4 \ldots W_{n-1}W_n$, where W_i is orthogonal and has the form

$$(4.16) \qquad \begin{bmatrix} 1 \ddots 1 & & \bigcirc \\ & \boxed{\begin{matrix} * & * & * \\ * & * & * \\ * & * & * \end{matrix}} & \\ \bigcirc & & 1 \ddots 1 \end{bmatrix}$$

with the center of the indicated 3×3 submatrix being on position i-1 of the diagonal. It can be checked that WR will have (possibly) nonzero elements in the two subdiagonals.

The above discussion is also presented in [2,3].

5. Mechanics of Updating QR

The process described here was first proposed by G. W. Stewart [12]. Let Q, R, y, and p be given, and denote the columns of Q by q_i $(i = 1,\ldots,n)$.

Step 1. Form $G = G_1 G_2 \ldots G_{n-1}$ as a product of Givens matrices chosen so that the vector

$$(5.1) \qquad p^T G_1 G_2 \ldots G_{i-1}$$

has zeros in positions 1, 2, ..., i-1 and G_i introduces a zero into the ith position. Then $RG = H$ is upper Hessenberg, and

$$(5.2) \qquad (QR + yp^T)G = QH + \|p\|\, ye_n^T ,$$

where e_n is the nth coordinate vector.

Step 2. Use the modified Gram-Schmidt process to find scalars $\tau_1, \tau_2, \ldots, \tau_n$ such that

$$(5.3) \qquad (\|p\|\, y) = \sum_{j=1}^{n} \tau_j q_j + r$$

where r is the residual vector. That is, if $t^T = [\tau_1, \ldots, \tau_n]$, then

$$(5.4) \qquad (\|p\|\, y) = Qt + r.$$

The residual vector r is in the orthogonal complement of the column space of Q. We may take the normalized vector $r/\|r\|$ to be an n+1st column q_{n+1} of Q, and we have:

$$(5.5)\qquad (QR+yp^T)G = \left[Q \,\vdots\, q_{n+1}\right]\left(H + te_n^T + \|r\|\, e_{n+1}e_n^T\right)$$
$$= \left[Q \,\vdots\, q_{n+1}\right]\tilde{H}\ .$$

The matrix $\tilde{H}$ is $(n+1) \times n$ right Hessenberg:

$$(5.6)\qquad \tilde{H} = \begin{bmatrix} \tilde{h} & \tilde{h} & \tilde{h} & (\tilde{h}+\tau) \\ \tilde{h} & \tilde{h} & \tilde{h} & (\tilde{h}+\tau) \\ 0 & \tilde{h} & \tilde{h} & (\tilde{h}+\tau) \\ 0 & 0 & \tilde{h} & (\tilde{h}+\tau) \\ 0 & 0 & 0 & \|r\| \end{bmatrix}\ .$$

We can remove the Givens transformations represented by G while leaving the right Hessenberg form of $\tilde{H}$ intact by carrying out further sequences of Givens transformations:

<u>Step 3</u>: For $i = n-1, n-2, \ldots, 1$ apply G_i on the right to $\tilde{H}$, which will introduce a (possibly) nonzero element into position $(i+2,i)$. Apply a Givens transformation M_i to $\tilde{H}$ on the left (and correspondingly to $[Q \,\vdots\, q_{n+1}]$ on the right) to zero out this new entry:

(5.7) 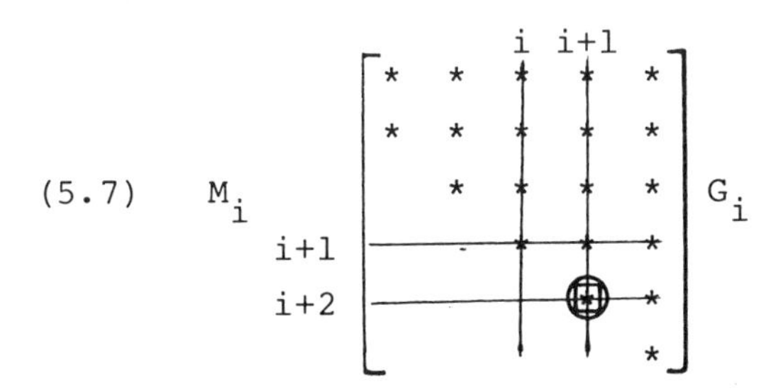

This results in the factorization

$$(5.8)\qquad \left([Q \,\vdots\, q_{n+1}]M^T\right)\left(M\tilde{H}G^T\right) = \hat{Q}\hat{H}$$

where $\hat{Q} = [Q \mid q_{n+1}]M^T$ is $m \times (n+1)$ and has orthonormal columns, and $\hat{H}$ is $(n+1) \times n$ and right Hessenberg.

Step 4: A final product of Givens transformations N is formed to reduce H to right triangular form:

$$(\hat{Q}N^T)(N\hat{H}) = (\hat{Q}N^T)\bar{R}. \tag{5.9}$$

The first n columns of $(\hat{Q}N^T)$ are to be taken as $\bar{Q}$; and indeed, $\bar{Q}\bar{R} = QR + yp^T$.

6. Monitoring and Efficiency

Both of the updating processes described require only $O(n[m+n])$ arithmetic operations to carry out, which makes them a much more efficient way of finding the factorizations in question than using a direct decomposition of J_k at each cycle.

The update computations have been constructed from such elemental steps as the multiplication of an orthogonal transformation onto a matrix, the solving of a triangular linear system, and the application of the modified Gram-Schmidt process--all of which have excellent round-off error properties.

A further advantage to maintaining J_k in some factored form is to be connected with the desire to monitor the "niceness" of the pseudo-Jacobian. It has been reported in practice (Wilkinson, Gill, Murray) that if a positive-definite matrix is decomposed into factors $\hat{R}^T\Delta\hat{R}$, where Δ is diagonal and $\hat{R}$ is unit right-triangular, then ill-conditioning in the positive definite matrix tends to reflect itself in Δ. Since the elements of Δ would correspond to the squares of the diagonal elements of our matrix R, this suggests that the observation of these numbers could be useful for monitoring the behavior of J.

Furthermore, in the case of the $R^T R$ factorization, the matrix J can be regarded with suspicion if the 2×2 principal minor of the matrix D in (4.7) is poorly conditioned.

These observations could be used as signals to apply pseudo-iterations, to expend energy in some way toward obtaining a better matrix J, or to study difficulties in the formulation of the problem itself.

7. Examples

Two experimental codes using the ideas presented here have been prepared and run on the CDC 6600/6400 at The University of Texas at Austin. The mix between pseudo-Newton and pseudo-steepest-descent steps was obtained from the Marquardt-Levenberg formula (2.5), where λ_k was set, simplistically, to be $||f(x_k)||$ and γ_k was chosen by means of a linear search (at an average cost of 3 function evaluations per cycle).

The diagram on the following page displays the number of function evaluations against the logarithm of $||f||^2$ in a minimization of the Powell 4-variable "singular" function using the QR code. The linear nature of the convergence is evident, and this graph is typical of those obtained for other test functions, though it must be admitted that the slope for this one is particularly steep.

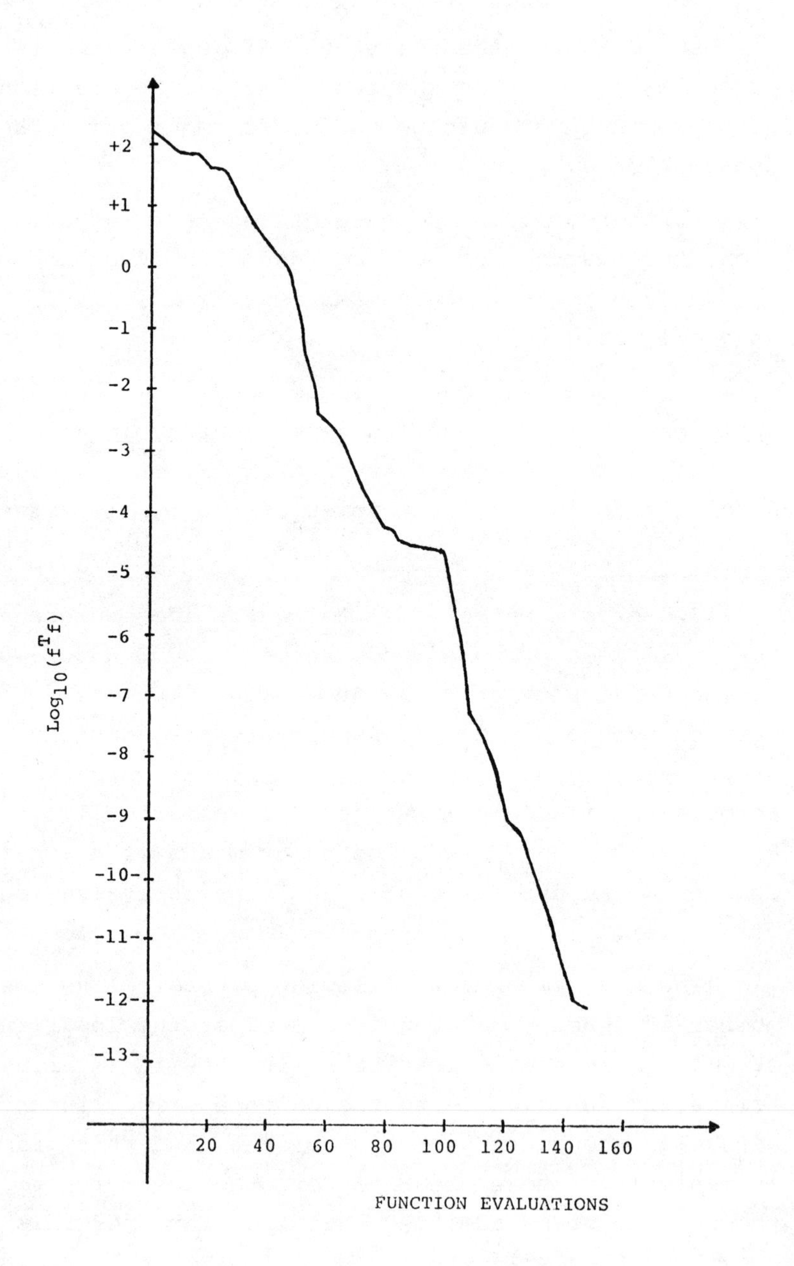
+2
+1
0
-1
-2
-3
-4
-5
-6
-7
-8
-9
-10-
-11-
-12-
-13-
$Log_{10}(f^Tf)$
20
40
60
80
100
120
140
160
FUNCTION EVALUATIONS

To indicate the evidence about the condition of J which might be read from the diagonal elements of R we present the following table produced from the minimization of $\|f(x)\|^2 = \|Jx\|^2$, where the matrix J (the Jacobian of f) was taken to be the principal $n \times n$ minor of the Hilbert matrix:

QR Code

n	final f^Tf	no. evaluations	r_{ii}^2
5	8.75×10^{-10}	109	$2.88 \times 10^{-6} \leftarrow r_{ii}^2 \rightarrow 1.48$
10	6.48×10^{-8}	151	$1.80 \times 10^{-5} \leftarrow r_{ii}^2 \rightarrow 1.69$

R^TR Code

n	final f^Tf	no. evaluations	r_{ii}^2
5	3.65×10^{-11}	109	$4.64 \times 10^{-7} \leftarrow r_{ii}^2 \rightarrow 1.59$
10	7.38×10^{-7}	151	$1.33 \times 10^{-7} \leftarrow r_{ii}^2 \rightarrow 1.56$

Note that the r_{ii}^2 range over some six orders of magnitude, reflecting the ill-condition of the problem.

Finally, to indicate that pseudo-Jacobian techniques can work in a practical situation, we present two graphs obtained from a free-knot least-squares spline data-fitting program built around the R^TR test code. For any given set of knots, one part of this code computes the best least squares fitting spline using a modified B-spline representation, and returns the residuals at the data points along with the representation. The R^TR program, in turn, regards the knots as variables and adjusts them to reduce the sum of squares of the residuals. The entire code is

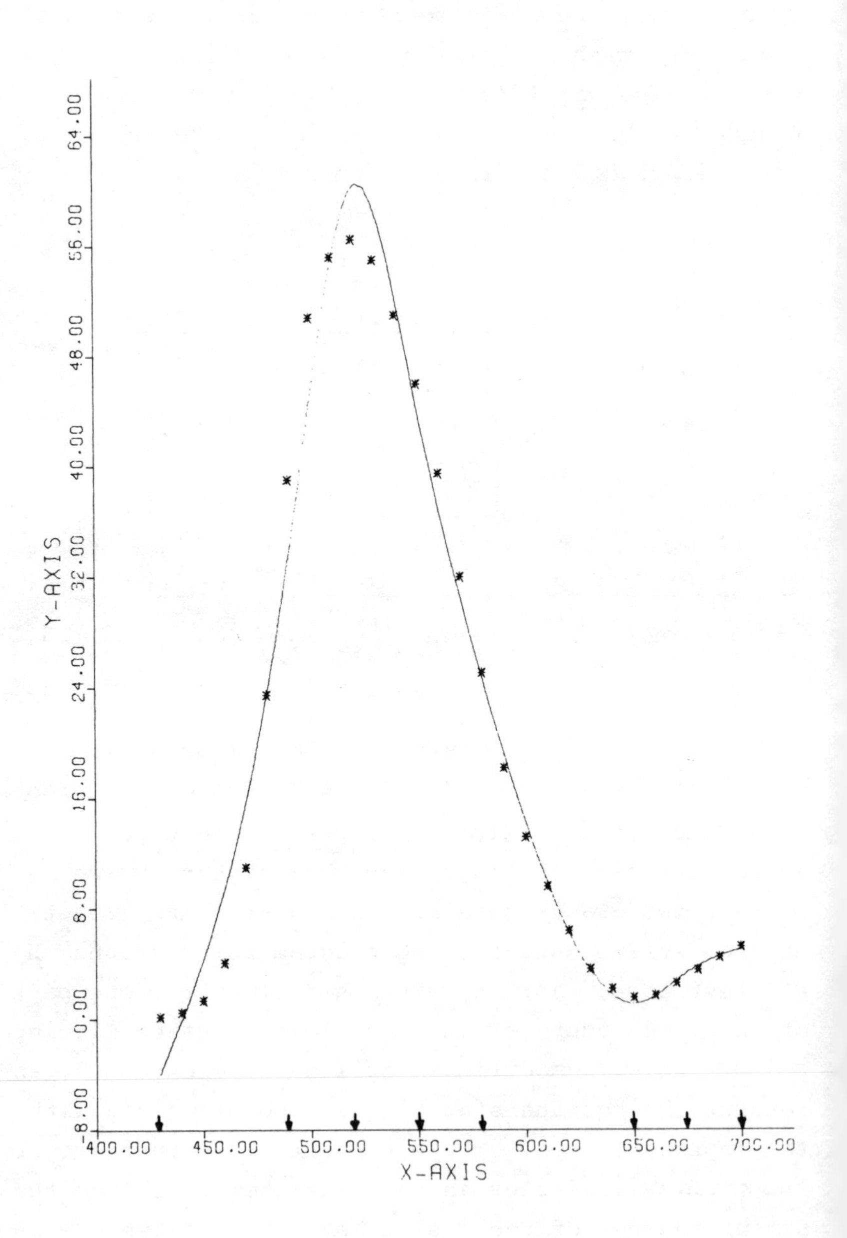
64.00
56.00
48.00
40.00
32.00
24.00
16.00
8.00
0.00
-8.00
Y-AXIS
400.00
450.00
500.00
550.00
600.00
650.00
700.00
X-AXIS

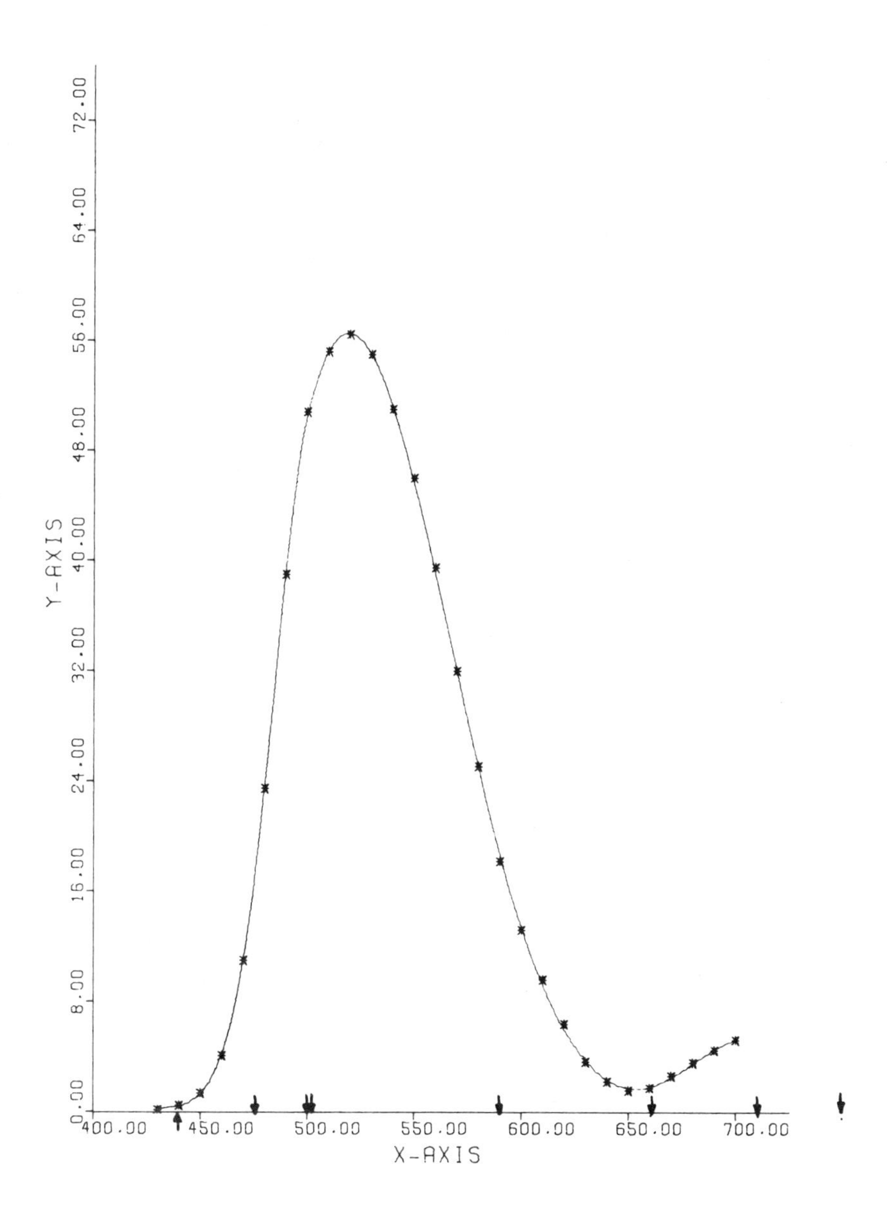
72.00
64.00
56.00
48.00
40.00
32.00
24.00
16.00
8.00
0.00
Y-AXIS
400.00
450.00
500.00
550.00
600.00
650.00
700.00
X-AXIS

designed to run interactively, permitting a user at a graphics terminal to select the starting knots. In the graphs below the positions of the knots are indicated by arrows and the data by asterisks. The first graph gives the starting knots and the resulting fit, the second shows the result after 80 function evaluations expended in minimization. Since 8 knots were used, this represents about the work of computing 9 numeric approximations to the Jacobian matrix. A commercial code based upon numeric derivatives was applied to the same problem and, after a great many more than 9 cycles, had still not succeeded in significantly improving the initial fit.

Bibliography

[1] Bartels, R. H. (1972) "Nonlinear Least Squares Without Derivatives: An Application of the QR Matrix Decomposition," Center for Numerical Analysis Report CNA-44, The University of Texas, Austin, Texas 78712.

[2] Bartels, R. H. (1973) "The Cholesky Factorization for Derivative-free Nonlinear Least Squares," Center for Numerical Analysis Technical Report CNA-64, The University of Texas, Austin, Texas 78712.

[3] Bartels, R. H. (1973) "A Cholesky-Factor Updating Technique for Rank-two Matrix Modifications," Center for Numerical Analysis Technical Report CNA-71, The University of Texas, Austin, Texas 78712.

[4] Bartels, R. H., Golub, G. H., and Saunders, M. A. (1970) "Numerical Techniques in Mathematical Programming," in *Nonlinear Programming*, edited by Rosen, Mangasarian, and Ritter, Academic Press, New York, pp. 123-176.

[5] Fletcher, R. (1968) "Generalized Inverse Methods for the Best Least Squares Solution of Systems of Nonlinear Equations," Comp. J. 10, pp. 392-399.

[6] Fletcher, R. (1970) "Generalized Inverses for Nonlinear Equations and Optimization," in Numerical Methods for Nonlinear Algebraic Equations edited by P. Rabinowitz, Gordon and Breach, New York, pp. 75-85.

[7] Gill, P. E. and Murray, W. (1973) "A Numerically Stable Form of the Simplex Algorithm," J. of Linear Algebra and Applics. 6, pp. 99-138.

[8] Gill, P. E., Golub, G. H., Murray, W. and Saunders, M. A. (1972) "Methods for Modifying Matrix Factorizations," Computer Science Department Technical Report STAN-CS-72-322, Stanford University, Stanford, California 94305.

[9] Golub, G. H. (1965) "Numerical Methods for Solving Linear Least Squares Problems," Numer. Math. 7, pp. 206-216.

[10] Osborne, M. R. (1972) "Some Aspects of Nonlinear Least Squares Calculations," in Numerical Methods for Nonlinear Optimization edited by F. A. Lootsma, Academic Press, New York, pp. 171-189.

[11] Powell, M.J.D. (1970) "A Hybrid Method for Nonlinear Equations," in Numerical Methods for Nonlinear Algebraic Equations edited by P. Rabinowitz, Gordon and Breach, New York, pp. 87-114.

[12] Stewart, G. W., Private communication.

NEWTON DERIVED METHODS FOR NONLINEAR EQUATIONS AND INEQUALITIES

by

E. Polak and I. Teodoru
University of California, Berkeley

1. Introduction

It has been shown by Pshenichnyi [6] and Robinson [7] that Newton's method can be extended to the solution of systems of equations and inequalities, where the number of variables may be larger than the number of equations and inequalities. These extensions of Newton's method converge quadratically when started from a good initial guess, but, just like Newton's method, may diverge when started from a poor initial guess. In a recent, as yet unpublished paper, Huang [2'] described a globally convergent modification of Robinson's algorithm.

Our paper presents an alternative stabilized version of Robinson's algorithm which has a greater rate of convergence and requires less strict assumptions than Huang's algorithm. In addition, we present an "iterated" version which is more efficient. The stabilization is accomplished by using an Armijo-type gradient method [1] until a battery of tests indicates that one is close enough to a solution for Robinson's

---

Research sponsored by the National Science Foundation Grant GK-37672 and the U.S. Army Research Office --Durham Contract DAHC04-73-C-0025.

extension of Newton's method to converge. Thus we obtain algorithms which are globally convergent and have root rate of convergence $r \in (1,2]$ depending on the choice of parameters. We show, following Brent [2], that there is a choice of parameters which maximizes the efficiency of the algorithm. Our computational experience has been most encouraging.

2. The Algorithm: Convergence

The algorithm we are about to state solves problems of the form: find $z \in \mathbb{R}^n$ such that

$$g(z) = 0 \qquad f(z) \leq 0 \tag{1}$$

where $g: \mathbb{R}^n \to \mathbb{R}^m$, $f: \mathbb{R}^n \to \mathbb{R}^q$ are continuously differentiable functions. We use superscripts to denote components of f, g, z, etc.

For some $b \gg 1$ and $\|\cdot\|$ denoting the Euclidean norm, we define $\bar{f}: \mathbb{R}^n \to \mathbb{R}^{q+1}$ by

$$\bar{f}^j(z) \triangleq \begin{cases} f^j(z) & j = 1,\ldots,q \\ \|z\|^2 - b & j = q+1 \end{cases} \tag{2}$$

We shall use the following notation:

$$h(z) \triangleq \begin{bmatrix} g(z) \\ \bar{f}(z) \end{bmatrix} \qquad H(z) \triangleq \frac{\partial h(z)}{\partial z} \tag{3}$$

$$G(z) \triangleq \frac{\partial g(z)}{\partial z}, \qquad \bar{F}(z) \triangleq \frac{\partial \bar{f}(z)}{\partial z} \tag{4}$$

$$\bar{f}^{j+}(z) \triangleq \max\{0, \bar{f}^j(z)\} \quad j=1,2,\ldots,q+1 \tag{5}$$

We also define a cost function $f^o: \mathbb{R}^n \to \mathbb{R}^1$ by

$$f^o(z) \triangleq \frac{1}{2}\|g(z)\|^2 + \frac{1}{2}\|\bar{f}^+(z)\|^2 \tag{6}$$

It is not difficult to see that f^o is continuously differentiable and

$$\nabla f^o(z) = G(z)^T g(z) + \bar{F}(z)^T \bar{f}^+(z) \qquad (7)$$

Finally, given any $z_0 \in \mathbb{R}^n$

$$C(z_0) \triangleq \{z | f^o(z) \le f^o(z_0)\} \qquad (8)$$

We now state assumptions which ensure that our algorithm is globally convergent.

Assumption 1: The derivative matrices $G(\cdot)$ and $F(\cdot)$ are Lipschitz continuous. ∎

Assumption 2: The pair $(\bar{F}(z), G(z))$ satisfies the Robinson LI condition [7] for all $z \in C(z_0)$, where z_0 is the initial guess to a solution for (1) and b is sufficiently large to ensure that the set $C(z_0)$ contains at least one such solution; i.e. for all $z \in C(z_0)$

$$u^T\bar{F}(z) + v^T G(z) = 0 \qquad (9)$$

and $u \geqq 0$, implies that $u = 0$ and $v = 0$. ∎

Algorithm

Data: $z_0 \in \mathbb{R}^n$, $b >> ||z_0||^2$, $\alpha \in (0,1/2)$, $\beta \in (0,1)$, $\hat{\ell} > 1$, $\gamma \in (0,1)$, $k \in N^+$.

Step 0: Set $i = 0$, $j = 0$, $s = 1$, $p = 0$.

Comment: i is the iteration index, the Jacobians $\bar{F}$ and G are evaluated at z_j, $s = i - j + 1$ is the number of times the same Jacobians have been used.

Step 1: Compute $g(z_i)$, $\bar{f}(z_i)$, $f^o(z_i)$, $G(z_j)$, and $\bar{F}(z_j)$. Stop if $f^o(z_i) = 0$.

Comment: When actually programming this algorithm, compute $G(z_j)$ and $\bar{F}(z_j)$ only as necessary (i.e. every k steps).

Step 2: Compute a vector v_i which solves the problem

$$\text{minimize}\{ \|v\|_\infty \mid g(z_i) + G(z_j)v = 0, \quad \bar{f}(z_i) + \bar{F}(z_j)v \leqq 0\} \tag{10}$$

where

$$\|v\|_\infty \triangleq \max_r |v^r| \tag{11}$$

Comment: Due to Assumption 2, the linearized problem (10) always has a solution, obtainable by linear programming techniques.

Step 3: If $\|v_i\|_\infty \leq \gamma^p$ and $z_i + v_i \in C(z_0)$, set $z_{i+1} = z_i + v_i$, set $p_i = p$, set $p = p + 1$, set $i = i + 1$ and go to step 14; else go to step 4.

Step 4: Set $w_i = v_i$, $\phi(z_i) = -2f^o(z_i)$.

Step 5: If $j = i$ go to step 11; else go to step 6.

Step 6: Set $\ell = 0$.

Step 7: Compute $f^o(z_i + \beta^\ell w_i)$.

Step 8: If

$$f^o(z_i + \beta^\ell w_i) - f^o(z_i) \leq \beta^\ell \alpha\phi(z_i) \tag{12}$$

set $\ell_i = \ell$, set $z_{i+1} = z_i + \beta^{\ell_i} w_i$, set $i = i + 1$ and go to step 14; else go to step 9.

Step 9: If $\ell < \hat{\ell}$ set $\ell = \ell + 1$ and go to step 7; else go to step 10.

Step 10: Compute $\nabla f^o(z_i)$, set $w_i = -\nabla f^o(z_i)$ and set $\phi(z_i) = -\|\nabla f^o(z_i)\|^2$.†

Step 11: Set $\ell = 0$.

Step 12: Compute $f^o(z_i + \beta^\ell w_i)$.

Step 13: If (12) is satisfied, set $\ell_i = \ell$, set $z_{i+1} = z_i + \beta^{\ell_i} w_i$, set $i = i + 1$ and go to step 14; else set $\ell = \ell + 1$ and go to step 12.

Step 14: If $s < k$, set $s = s + 1$ and go to step 1; else, set $s = 1$, $j = i$ and go to step 1.

We shall now establish the convergence properties of the algorithm in four stages. First we shall show that if $\{z_i\}_{i=0}^{\infty}$ is constructed by the algorithm, then it contains a subsequence which converges to a solution. Next we shall show that the relation $z_{i+1} = z_i + v_i$, i.e. $\|v_i\|_\infty \leq \gamma^{p_i}$, $z_i + v_i \in C(z_0)$, holds an infinite number of times. Next we shall show that if there is an i' such that the test $\|v_i\|_\infty \leq \gamma^{p_i}$, $z_i + v_i \in C(z_0)$ in step 3 is satisfied for all $i \geq i'$ and $\|v_i\|$ is sufficiently small, then $z_i \to \hat{z}$, a solution of (1). This result takes the form of a local convergence theorem. We shall

†When the solution of (10) is not costly, replace step 10 with: Compute $G(z_i)$, $F(z_i)$, set $j = i$, set $s = 1$ and go to step 2.

complete the proof by exhibiting the existence of an i' such that $z_{i+1} = z_i + v_i$ for all $i \geq i'$.

Proposition 1: Let v_i be computed by the algorithm at z_i, from (10).

Then $\phi(z_i) = 0$ if and only if z_i solves (1). ∎

Lemma 1: Suppose that $z \in \mathbb{R}^n$ is such that $\phi(z) < 0$, where $\phi(z)$ was defined in the algorithm (step 4 or step 10). Then there exists an integer $\bar{\ell} \geq 0$, finite, and an $\varepsilon(z) > 0$ such that

$$f^o(z' + \beta^{\bar{\ell}}w') - f^o(z') \leq \beta^{\bar{\ell}}\alpha\phi(z') \tag{13}$$

for all $z' \in B(z,\varepsilon(z)) \triangleq \{z' \mid \|z' - z\| \leq \varepsilon(z)\}$, for all w' satisfying for some $Q < \infty$

$$\|w'\| \leq Q \quad \text{and} \quad \langle\nabla f^o(z'),w'\rangle \leq \phi(z') \tag{14}$$

Proof: Because of Assumption 1, $\nabla f^o(\cdot)$ is continuous. Hence $\phi(\cdot)$ is continuous and there exists an $\varepsilon(z) > 0$ such that $\phi(z') \leq \phi(z)/2 < 0$ for all $z' \in B(z,\varepsilon(z))$. Since $B(z,\varepsilon)z))$ is compact, $\nabla f^o(\cdot)$ is uniformly Lipschitz continuous on this set, with constant L, say, and hence, for any $\lambda > 0$, and w' satisfying (14),

$$\begin{aligned} &f^o(z'+\lambda w') - f^o(z') - \lambda\langle\nabla f^o(z'),w'\rangle \\ &= \int_0^1 \langle\nabla f^o(z'+s\lambda w') - \nabla f^o(z'),\lambda w'\rangle ds \leq \frac{L\lambda^2}{2}\|w'\|^2 \end{aligned} \tag{15}$$

Consequently,

$$\begin{aligned} f^o(z'+\lambda w') - f^o(z') &\leq \lambda\alpha\langle\nabla f^o(z'),w'\rangle + \\ &\quad \lambda\left[(1-\alpha)\langle\nabla f^o(z'),w'\rangle + \frac{L\lambda}{2}\|w'\|^2\right] \\ &\leq \lambda\alpha\phi(z') + \frac{\lambda}{2}[(1-\alpha)\phi(z) + L\lambda Q^2] \end{aligned} \tag{16}$$

Let $\bar{\ell} \geq 0$ be the smallest integer such that $(1-\alpha)\ \phi(z) + L\beta^{\bar{\ell}}Q^2 \leq 0$. Then, clearly $\bar{\ell}$ satisfies (13) for all $z' \in B(z,\varepsilon(z))$, for all w' satisfying (14) and the lemma is proved.

Corollary 1: The algorithm is well defined, i.e., it does not jam up between steps 12 and 13.

Proof: We know from Proposition 1 that it is not possible to reach step 11 with $\phi(z_i) < 0$ and $w_i = 0$. Next, since $f^o(z)$ contains the term $((\|z\|^2 - b)^+)^2$, it is clear that $C(z_0)$ is compact and hence there exists an $M < \infty$ such that $\max\{\|\nabla f^o(z)\|_\infty, \|h(z)\|_\infty\} \leq M$ for all $z \in C(z_0)$. Hence, making use of (56) and the equivalence of norms in $\mathbb{R}^n$, we conclude that there exists a $Q \in [M,\infty)$, such that $\|w_i\| \leq Q$ for any z_i constructed by the algorithm. Next, if $w_i = v_i$ we have from (10) that $< \nabla f^o(z_i), w_i > \leq -2f^o(z_i) = \phi(z_i)$, whereas if $w_i = -\nabla f^o(z_i)$ the same result follows from the definition in Step 10. Hence by Lemma 1 there exists an $\ell_i < \infty$ satisfying (12). ∎

Lemma 2: Suppose that the algorithm has constructed an infinite sequence $\{z_i\}_{i=0}^{\infty}$. Then $\{z_i\}_{i=0}^{\infty}$ has at least one accumulation point $\hat{z}$ which solves (1), i.e. $g(\hat{z}) = 0$, $\bar{f}(\hat{z}) \leq 0$.

Proof: First, suppose that there is an $\hat{i}$ such that for all $i \geq \hat{i}$, z_{i+1} is constructed according to the formula in steps 8 or 13 $(z_{i+1} = z_i + \beta^{\ell_i}w_i)$. Then it follows directly from Lemma 1, and theorem 1.3.9 in [5] that every accumulation point $\hat{z}$ of $\{z_i\}_{i=0}^{\infty}$ satisfies $\nabla f^o(\hat{z}) = 0$ and hence, from Assumption 2, is a solution of (1). Furthermore, since the set $C(z_{\hat{i}}) = \{z | f^o(z) \leq f^o(z_{\hat{i}})\}$ is compact, and $z_i \in C(z_{\hat{i}})$ for all $i \geq \hat{i}$, it follows that $\{z_i\}$ has at least one accumulation point $\hat{z}$ which solves (1).

Next, suppose there is no $\hat{i}$ such that $z_{i+1} = z_i + \beta^{\ell_i} w_i$ for all $i \geq \hat{i}$. Then there must exist an infinite subset $K \subset \{0,1,2,...\}$ such that $\|v_i\|_\infty \leq \gamma^{p_i}$ for all $i \in K$, and $p_i \to \infty$ as $i \to \infty$, $i \in K$. Consequently, $v_i \to 0$ as $i \to \infty$, $i \in K$. This implies that $g(z_i) \to 0$, $\bar{f}(z_i) \to \hat{f} \leqq 0$ as $i \to \infty$, $i \in K$ since the Jacobians $G(\cdot)$ and $\bar{F}(\cdot)$ are bounded over $C(z_0)$. Since $\{z_i\}_{i \in K}$ is compact, there exists an infinite subset $K' \subset K$ and a $\hat{z} \in C(z_0)$ such that $z_i \to \hat{z}$ as $i \to \infty$, $i \in K'$, and therefore $g(\hat{z}) = 0$, $\bar{f}(\hat{z}) = \hat{f} \leqq 0$, i.e. $\{z_i\}_{i=0}^{\infty}$ has an accumulation point $\hat{z}$ which solves (1).

<u>Corollary 2</u>: Suppose that $\{z_i\}_{i=0}^{\infty}$ is a sequence constructed by the algorithm and suppose that $\hat{z}$ is an accumulation point of $\{z_i\}_{i=0}^{\infty}$ which solves (1). If $K \subset \{0,1,2,...\}$ is an index set identifying a subsequence of $\{z_i\}_{i=0}^{\infty}$ which converges to $\hat{z}$, i.e. $z_i \to \hat{z}$ as $i \to \infty$, $i \in K$, then $v_i \to 0$ as $i \to \infty$, $i \in K$.

Proof: Note that

$$\|v_i\|_\infty \le \nu_i \triangleq \min\left\{\|v\|_\infty \mid g(z_i) + G(z_j)v = 0,\ \bar{f}^+(z_i) + \bar{F}(z_j)v \leqq 0\right\} \tag{17}$$

and $\nu_i \to 0$ as $i \to \infty$, $i \in K$ by (56) and Lemma 2. Hence $v_i \to 0$ as $i \to \infty$, $i \in K$. ■

Corollary 3: Suppose that $\{z_i\}_{i=0}^{\infty}$ is a sequence constructed by the algorithm, and suppose that $\{z_i\}_{i \in K}$ is a subsequence converging to a solution $\hat{z}$ of (1), where $K \subset \{0,1,2,...\}$. Then there is an infinite subset $K' \subset K$ such that $z_{i+1} = z_i + v_i$ for all $i \in K'$.

Proof: By Corollary 2, $v_i \to 0$ as $i \to \infty$, $i \in K$. Hence, since $f^o(z_i) \to 0$ as $i \to \infty$, $i \in K$, and $f^o(\cdot)$ is uniformly continuous on $\{z_i\}_{i \in K}$, there exists an integer i" such that for all $i \in K$, $i \ge i"$, $\|v_i\|_\infty \le \gamma$,

$$f^o(z_i) \le \frac{1}{2} f^o(z_0) \tag{18}$$

$$f^o(z_i + v_i) - f^o(z_i) \le \frac{1}{2} f^o(z_0) \tag{19}$$

i.e., $z_i + v_i \in C(z_0)$ for all $i \ge i"$, $i \in K$; moreover the construction $z_{i+1} = z_i + v_i$, $i \in K$, must occur at least once, so that $p_{i"}$ is well defined. Now let p_i, $i \in K$, $i \ge i"$ be an arbitrary integer. Then, since $v_i \to 0$ as $i \to \infty$, $i \in K$, there exists a j such that $i + j \in K$ and $\|v_{i+j}\|_\infty \le \gamma^{p_i+1}$, which together with (18), (19) implies that $z_{i+j+1} = z_{i+j} + v_{i+j}$. Thus there exists an infinite subset

$K' \subset K$ such that $z_{i+1} = z_i + v_i$ for all $i \in K'$. ■

<u>Theorem 1 (Local Convergence)</u>: Let $g: \mathbb{R}^n \to \mathbb{R}^m$ and $f: \mathbb{R}^n \to \mathbb{R}^q$ be differentiable functions, with $G(\cdot)$, $\bar{F}(\cdot)$ uniformly Lipschitz continuous with constant L. Let $b > 0$, $\gamma \in (0,1)$ and an integer $k \geq 1$ be given. Suppose there exists a $y_0 \in \mathbb{R}^n$ such that the pair $[\bar{F}(y_0), G(y_0)]$ satisfies the LI condition and (see also (58))

$$h \triangleq \mu^* L \eta \leq \gamma - \gamma^2, \quad \mu^* \geq \sup_{y \in Y} \mu[\bar{F}(y), G(y)]$$

$$Y = \left\{ y \mid \|y - y_0\|_\infty \leq \frac{\eta}{1-\delta} \right\}, \quad \delta = \frac{1}{2} - \sqrt{\frac{1}{4} - h} \tag{20}$$

and

$$\eta \triangleq \min\{ \|v\|_\infty \mid g(y_0) + G(y_0)\, v = 0, \quad \bar{f}(y_0) + \bar{F}(y_0)v \leqq 0\} \tag{21}$$

For $i = 0,1,2,\ldots$, let $j(i) \triangleq k[i/k]$, where $[i/k]$ denotes the integer part of i/k. Then the iterative process

$$y_{i+1} \in \text{Arg min}\{ \|y - y_i\|_\infty \mid g(y_i) + G(y_{j(i)})(y-y_i) = 0, \quad \bar{f}(y_i) + \bar{F}(y_{j(i)})(y-y_i) \leqq 0\} \; i = 0,1,2,\ldots, \tag{22}$$

results in well defined sequences $\{y_i\}_{i=0}^{\infty}$ such that any such sequence converges to some $\hat{y}$ satisfying $g(\hat{y}) = 0$, $\bar{f}(\hat{y}) \leqq 0$.

<u>Proof</u>: First suppose that the process (22) results in a well defined sequence $\{y_i\}_{i=0}^{\infty} \subset Y$. For $i = 0,1,2,\ldots$, consider the following associated linear system:

$$g(y_i) + G(y_{j(i)})(y-y_i) - [g(y_{i-1}) + G(y_{j(i-1)})(y_i-y_{i-1})] = 0 \tag{23}$$

$$\bar{f}(y_i) + \bar{F}(y_{j(i)})(y-y_i) - [\bar{f}(y_{i-1}) + \bar{F}(y_{j(i-1)})(y_i-y_{i-1})] \leq 0 \tag{24}$$

Since, by inspection, any solution to (23) - (24) is a feasible point for (22), we obtain the following bound from (3), (56) and the Lagrange formula:

$$\begin{aligned} \|y_{i+1}-y_i\|_\infty &\leq \mu_{j(i)} \|h(y_i) - h(y_{i-1}) - H(y_{j(i-1)})(y_i-y_{i-1})\|_\infty \\ &\leq \mu_{j(i)} \int_0^1 \|H(y_{i-1}+s(y_i-y_{i-1})) - H(y_{j(i-1)})\|_\infty \, ds \, \|y_i-y_{i-1}\|_\infty \\ &\leq \mu_{j(i)} L \left[\|y_{i-1}-y_{j(i-1)}\|_\infty + \|y_i-y_{i-1}\|_\infty\right] \|y_i-y_{i-1}\|_\infty \\ &\leq \mu_{j(i)} L \left[\sum_{\nu=j(i-1)}^{i-1} \|y_{\nu+1}-y_\nu\|_\infty\right] \|y_i-y_{i-1}\|_\infty \end{aligned} \tag{25}$$

Next, from (20) it follows that δ is real for any $\gamma \in (0,1)$ and $h = \delta - \delta^2 \leq \delta \leq \gamma$. (26)

We shall now show by induction that the process (22) is well defined for $i = 0,1,2,\ldots$, that $y_i \in Y$ for $i = 0,1,2,\ldots$, and that

$$\|y_{i+1}-y_i\|_\infty \le \delta \|y_i-y_{i-1}\|_\infty, \; i = 1,2,\ldots \tag{27}$$

Note that a y_1 satisfying (22) for $i = 0$ exists by hypothesis, and that $y_1 \in Y$. Next, let $i = 1$. If $k > 1$, then $j(1) = 0$ and, by hypothesis, the Jacobians satisfy the LI condition at y_0. Consequently, by Theorem 5, there is at least one y_2 satisfying (22) for $i = 1$. If, on the other hand, $k = 1$, then because of the Lipschitz continuity of the Jacobians we have

$$\mu_0 \|H(y_1) - H(y_0)\|_\infty \le \mu_0 L \|y_1-y_0\|_\infty \le \mu^* L\eta = h < 1 \tag{28}$$

and hence by Theorem 5, the LI condition is satisfied for $i = 1$ and there exists at least one y_2 satisfying (22) for $i = 1$. It now follows from (25) that, whether $k = 1$ or $k > 1$,

$$\|y_2-y_1\|_\infty \le \mu_{j(1)} L[\|y_1-y_0\|_\infty] \|y_1-y_0\|_\infty \le \delta \|y_1-y_0\|_\infty \tag{29}$$

the last part of (29) because of (20) and (26). Also, $\|y_2-y_0\|_\infty \le \|y_2-y_1\|_\infty + \|y_1-y_0\|_\infty \le (1+\delta)\eta \le \frac{1}{1-\delta}\eta$, i.e., $y_2 \in Y$. Thus the sequence is well defined for $i = 0,1,2$, it is contained in Y, and (27) holds for $i = 1$. We proceed with the inductive step: for $i = N + 1$, either $j(N+1) = j(N)$ or $j(N+1) = N + 1$. If $j(N+1) = j(N)$, then, since the matrices in the linear programming problem (22) satisfy the LI condition at $i = N$, they also satisfy it at $i = N+1$ and y_{N+2} is well defined. If $j(N+1) \ne j(N)$, then, because of the uniform Lipschitz continuity of the Jacobians,

$$\mu_{j(N)} \|H(y_{N+1}) - H(y_{j(N)})\|_\infty \le \mu_{j(N)} L\|y_{N+1} - y_{j(N)}\|_\infty$$

$$\le \mu^* L \sum_{\nu=j(N)}^{N} \|y_{\nu+1}-y_\nu\|_\infty \le \mu^* L\eta \sum_{\nu=j(N)}^{N} \delta^\nu$$

$$\le h/(1-\delta) = \delta. \tag{30}$$

where the last line follows from (26). Hence by Theorem 5, the LI condition is satisfied for $i = N+1$ and thus y_{N+2} is well defined. It now follows from (25) and (26) that

$$\|y_{N+2} - y_{N+1}\|_\infty \le \delta\|y_{N+1} - y_N\|_\infty \tag{31}$$

i.e., (27) holds for $i = N+1$. Thus, the sequence $\{y_i\}_{i=0}^\infty$ is well defined, it satisfies (27) and is contained in Y. It must converge to a $\hat{y}$ because (27) holds for all i. It now follows from the continuity of g, G, $\bar{f}$ and $\bar{F}$, and the fact that $v_i = (y_{i+1}-y_i)$ converges to zero as $i \to \infty$, that $g(\hat{y}) = 0$, $\bar{f}(\hat{y}) \leqq 0$. This completes our proof. ∎

<u>Theorem 2</u>: Suppose that b in the algorithm is sufficiently large so that $\{z|g(z) = 0, \bar{f}(z) \leqq 0\} \neq \phi$, and suppose that Assumptions 1 and 2 are satisfied. Suppose that the algorithm has constructed an infinite sequence $\{z_i\}_{i=0}^\infty$. Then (i) there exists an $N \ge 0$ such that $z_{i+1} = z_i + v_i$ for all $i \ge N$ and (ii) $z_i \to \hat{z}$ as $i \to \infty$ with $g(\hat{z}) = 0$, $f(\hat{z}) \leqq 0$.

<u>Proof</u>: According to Lemma 2, there exists an infinite sequence $\{z_i\}_{i\in K}$ such that $z_i \to \hat{z}$, and

$f^o(z_i) \to f^o(\hat{z}) = 0$, as $i \to \infty$, $i \in K$, with $\hat{z}$ a solution of (1). By corollaries 2 and 3 there is an infinite subset $K' \subset K$, such that $v_i \to 0$ as $i \to \infty$, $i \in K'$ and $z_{i+1} = z_i + v_i$, for all $i \in K'$.

Next, let $\varepsilon_1 = f^o(z_0)/2$. Then, since $f^o(\cdot)$ is uniformly continuous on the compact set $C(z_0)$, there exists a $\delta_1 > 0$ such that

$$|f^o(z) - f^o(z')| < \varepsilon_1 \tag{32}$$

for all $||z' - z||_\infty \le \delta_1$. Now, since $f^o(z_i) \to 0$, $v_i \to 0$ as $i \to \infty$, $i \in K'$, there exists an $i' \in K'$ such that

$$f^o(z_{i'}) \le \varepsilon_1, \quad ||v_{i'}||_\infty \le (\gamma - \gamma^2)/\mu^* L \tag{33}$$

and

$$\omega \triangleq \frac{||v_{i'}||_\infty}{\frac{1}{2} + \sqrt{\frac{1}{4} - \mu^* L\, ||v_{i'}||_\infty}} \le \delta_1 \tag{34}$$

Then the conditions of Theorem 1 are satisfied at $y_0 = z_{i'}$ and any sequence $\{y_j\}_{j=0}^\infty$ constructed according to (22) must converge to a $\hat{y}$, which solves (1). Furthermore, from (27), $||y_j - y_0||_\infty \le \omega \le \delta_1$ for all $j = 1,2,\ldots$, and hence, from (32) and (33),

$$f^o(y_j) \le f^o(y_0) + \varepsilon_1 \le f^o(z_0) \tag{35}$$

i.e. $y_j \in C(z_0)$ for $j = 0,1,2,\ldots$. Hence, we must have $z_{i'+1} = y_1$. Now making use of (27) and (26) we obtain

$$\begin{aligned} ||v_{i'+1}||_\infty = ||y_2 - y_1||_\infty &\le \delta ||v_{i'}||_\infty \\ &\le \gamma ||v_{i'}||_\infty \le \gamma^{p_{i'}+1} = \gamma^{p_{i'+1}} \end{aligned} \tag{36}$$

and hence $z_{i'+2} = y_2$. Proceeding by induction, we now conclude that $z_{i'+j} = y_j$, $j = 0,1,2,\ldots$, and consequently $z_i \to \hat{y}$ a solution of (1). This completes our proof. ∎

3. Rate of Convergence and Efficiency

We derive now the root rate of convergence of the algorithm in Section 2 (see [4] sec. 9.1 for a definition and discussion of root rate). We shall show that the root rate of the iterative process (22), and hence of the process defined in Section 2, is $\geq r \triangleq \sqrt[k]{k+1}$. The proof will proceed in three steps. We shall first show that the sequence $\{s_{ik}\}_{i=0}^{\infty}$ of step lengths ($s_{ik} \triangleq \|y_{ik+1} - y_{ik}\|_\infty$) associated with the k-step process obtained from (22) has an R-rate $\geq k+1$. (Compare with results of Traub [10] and Shamanskii [9] for multistep methods.) Next we shall obtain a relation between the rate of convergence of the subsequence $\{s_{ik}\}_{i=0}^{\infty}$ and that of the sequence $\{s_i\}_{i=0}^{\infty}$. Finally, we shall show that the R-rate of the process (22) is the same as the R-rate of the sequence $\{s_i\}_{i=0}^{\infty}$.

Define

$$s_i \triangleq \|y_{i+1} - y_i\|_\infty \tag{37}$$

Then we have the following

Lemma 3: Let $g: \mathbb{R}^n \to \mathbb{R}^m$ and $f: \mathbb{R}^n \to \mathbb{R}^q$ be differentiable functions, with $G(\cdot)$ and $\bar{F}(\cdot)$ uniformly Lipschitz continuous with constant L. Let $b > 0$, $\gamma \in (0,1)$, $k \in N^+$ be given. Suppose there exists a $y_0 \in \mathbb{R}^n$ such that the pair $[\bar{F}(y_0), G(y_0)]$ satisfies the LI condition and

$$s_0 < \min\left\{\frac{1}{[a_k(\mu^*L)^k]^{k+1}}, \frac{\gamma-\gamma^2}{\mu^*L}, 1\right\}, \tag{38}$$

where a_k is defined by the recursive relation

$$a_{\ell+1} = a_\ell \sum_{i=0}^{\ell} a_i, \quad a_0 = 1 \quad \ell = 0,1,\ldots,k-1, \tag{39}$$

and μ^* satisfies (20).

(The existence of a y_1 solving (22) is ensured by the hypothesis).

Under these conditions, the sequence $\{s_i\}_{i=0}^{\infty}$ associated with the process (22) converges to 0 with root rate of at least $r \triangleq \sqrt[k]{k+1}$.[†]

Proof: From the local convergence theorem, the iterative process (22) constructs a well defined sequence $\{y_i\}_{i=0}^{\infty}$. Moreover, this sequence is Cauchy, and hence the sequence $\{s_i\}_{i=0}^{\infty}$ converges to 0.

We begin by showing that for any $\ell \in \{1,2,\ldots,k\}$ and for any $i = 0,1,\ldots,$ we can bound $s_{j(i)+\ell}$ by

$$s_{j(i)+\ell} \le a_\ell(\mu^*L)^\ell \, s_{j(i)}^{\ell+1} \tag{40}$$

with a_ℓ as above. The proof will be by induction on ℓ. From (25) we obtain, by replacing i with $j(i) + 1$ and using (20) and the definition (37):

[†]By theorem 9.2.7 in [4], r is the root rate of the sequence $\{s_i\}_{i=0}^{\infty}$ only if $0 < \overline{\lim}_{i\to\infty} s_i^{\frac{1}{r^i}} < 1$.

$$s_{j(i)+1} \leq \mu^* L\, s_{j(i)}^2 \tag{41}$$

which proves (40) for the case when $\ell = 1$. Assume next that (40) holds for $\ell = 1,2,\ldots,\bar{k}$, with $\bar{k} \leq k - 1$. Making again use of (25) and the inductive hypothesis, we obtain successively:

$$\begin{aligned} s_{j(i)+\bar{k}+1} &\leq \mu^* L \left(\sum_{\nu=j(i)}^{j(i)+\bar{k}} s_\nu \right) s_{j(i)+\bar{k}} \\ &= a_{\bar{k}} (\mu^* L)^{\bar{k}+1} s_{j(i)}^{\bar{k}+2} \sum_{\nu=0}^{\bar{k}} a_\nu (\mu^* L\, s_{j(i)})^\nu \\ &\leq a_{\bar{k}+1} (\mu^* L)^{\bar{k}+1} s_{j(i)}^{\bar{k}+2} \end{aligned} \tag{42}$$

where the last line follows from (38) and the fact that the sequence $\{s_i\}_{i=0}^{\infty}$ is a monotone decreasing sequence. The bound in (40) thus holds for any $\ell \in \{1,2,\ldots,k\}$ and, since i was arbitrary, it also holds for any $i \in N^+$.

A simple calculation will lead now to the desired result. Noting that $s_{j(i)+k} = s_{j(i+k)}$ and applying repeatedly (40) with $\ell = k$, we obtain

$$s_{j(i+k)} \leq a_k (\mu^* L)^k s_{j(i)}^{k+1} \leq \left\{ a_k (\mu^* L)^k \right\}^{\sum_{\nu=0}^{j(i)/k} (k+1)^\nu} s_0^{r^{j(i+k)}}. \tag{43}$$

Observe next that for any $i \in N^+$ and for any integer $k \geq 1$,

$$\sum_{\nu=0}^{\frac{j(i)}{k}} (k+1)^{\nu} = \frac{(k+1)^{\frac{j(i)}{k}+1} - 1}{k} < (k+1)^{\frac{j(i)}{k}+2} \qquad (44)$$

Without loss of generality, we can assume that $a_k(\mu^* L)^k \geq 1$ (since if it is not, we can choose a larger Lipschitz constant). We can then bound (43) using the inequality (44) by:

$$s_{j(i+k)} \leq \left\{a_k(\mu^* L)^k\right\}^{(k+1)^{(\frac{j(i)}{k}+2)}} s_0^{r^{j(i+k)}}$$

$$= \left\{[a_k(\mu^* L)^k]^{k+1} s_0\right\}^{r^{j(i+k)}} \triangleq \theta^{r^{j(i+k)}} \qquad (45)$$

Since we can express any $i = 1,2,\ldots$ as $i = j(i-1) + \ell$ for some $\ell \in \{1,2,\ldots,k\}$, we can combine (40) and (45) to obtain:

$$\overline{\lim_{i\to\infty}} \, s_i^{\frac{1}{r^i}} \leq \overline{\lim_{i\to\infty}} \left[a_\ell(\mu^* L)^\ell \, s_{j(i-1)}^{\ell+1}\right]^{\frac{1}{r^i}}$$

$$\leq \overline{\lim_{i\to\infty}} \, \theta^{\frac{r^{j(i-1)}(\ell+1)}{r^{j(i-1)} r^\ell}} = \theta^{\frac{\ell+1}{r^\ell}} < 1. \qquad (46)$$

Hence the sequence $\{s_i\}_{i=0}^{\infty}$ has a root rate of convergence $\geq r$ (by Theorem 9.2.7 in [4]), and this completes the proof of the lemma. ■

<u>Corollary 4</u>: Under the conditions of the preceding lemma, the following bound holds for s_i

$$s_i \leq A\,\theta^{\lambda r^i} \quad \text{for} \quad i = 1,2,\ldots \qquad (47)$$

where $A \triangleq a_k(\mu^* L)^k$ and

$$\lambda \triangleq \min\left\{\frac{\ell+1}{r^\ell} \mid \ell \in \{1,2,\ldots,k\}\right\} < 1. \quad \blacksquare$$

<u>Theorem 3:</u> Assume the conditions of Lemma 3 hold. Then the root rate of any sequence defined by (22) is at least $r = \sqrt[k]{k+1}$.

<u>Proof</u>: According to Theorem 9.2.7 in [4], we only need to show that

$$\overline{\lim_{i\to\infty}} \|y_i - \hat{y}\|_\infty^{\frac{1}{r^i}} < 1 \tag{48}$$

Obviously,

$$e_i \triangleq \|y_i - \hat{y}\|_\infty \le \sum_{\nu=i}^{\infty} s_\nu \le \sum_{\nu=i}^{\infty} A\,\theta^{\lambda r^\nu}$$

$$= A\,\theta^{\lambda r^i} \sum_{\nu=i}^{\infty} \theta^{\lambda(r^\nu - r^i)}$$

$$= A\,\theta^{\lambda r^i} \sum_{\nu=0}^{\infty} \theta^{\lambda r^i (r^\nu - 1)} \tag{49}$$

The series appearing in the last line of (49) converges for any $i \in N^+$, since it satisfies the root test for series. Moreover, denoting the infinite series in (49) by b_i, we observe that $\{b_i\}_{i=0}^{\infty}$ is a decreasing sequence (since $\theta^{\lambda r^{i+1}} < \theta^{\lambda r^i}$). Therefore, the bound in (49) becomes:

$$e_i \le A\, b_0\, \theta^{\lambda r^i} \tag{50}$$

which is equivalent to (48). $\blacksquare$

Thus we have proved the following:

<u>Theorem 4</u>: Suppose that b in the algorithm is sufficiently large so that $\{z | g(z) = 0, \bar{f}(z) \leq 0\} \neq \phi$, and suppose that Assumptions 1 and 2 are satisfied. Suppose that the algorithm has constructed an infinite sequence $\{z_i\}_{i=0}^{\infty}$. Then (i) $z_i \to \hat{z}$ as $i \to \infty$, with $g(\hat{z}) = 0$, $\bar{f}(\hat{z}) \leq 0$,

and (ii) $\overline{\lim_{i\to\infty}} \|z_i - \hat{z}\|_\infty^{\frac{1}{r^i}} < 1$ where $r \triangleq \sqrt[k]{k+1}$. ∎

We now show how to choose an optimal value of k. Brent [2] defined the efficiency of an algorithm as:

$$E \triangleq \frac{\ell n \ r}{w} \tag{51}$$

where r is the root rate of convergence of the algorithm and w is the average amount of work per iteration (e.g. number of function evaluations, CPU time, etc).

For the algorithm is Section 2 we have, by Theorem 4

$$E_k = \frac{\ell n \ (k+1)}{k \ w(k)} \tag{52}$$

It can be easily seen that using any reasonable definition for w, E_k will attain a maximum value for some $k \in N^+$. The value of the maximizer, k_{opt}, can be obtained either by experimentally evaluating w(k), or by making use of an explicit expression for w(k) when available. For example, if the number of function evaluations is the dominant factor in an iteration (as is the case in boundary value problems), then E_k has the form:

$$E_k = \frac{n(k+1)}{k\frac{(m+q)(n+k)}{k}} = \frac{n(k+1)}{(m+q)(n+k)} \tag{53}$$

and the vaue of k_{opt} can be determined a priori.

Conclusion

A limited amount of experimental evidence indicates that the algorithm works well even when the LI condition, which is sufficient to ensure global convergence of the algorithm, is not satisfied at all points of $C(z_0)$.

Finally, it should be observed that the algorithm can be easily adapted for solving boundary value problems with ordinary differential equations.

Appendix

We present below a summary of the results concerning the solution of mixed systems of equations and inequalities which have been used in this paper.

Consider the system

$$Az \leq b \qquad Cz = d \tag{54}$$

where $A \in \mathbb{R}^{(q+1)\times n}$, $C \in \mathbb{R}^{m\times n}$, $b \in \mathbb{R}^{q+1}$, $d \in \mathbb{R}^m$, and let $\mathcal{F}$ be the set of right-hand side vectors for which (54) has at least one solution:

$$\mathcal{F} \triangleq \{\binom{b}{d} \mid \exists\ z \in \mathbb{R}^n \ \text{s.t.}\ Az \leq b,\ Cz = d\} \tag{55}$$

It has been shown (e.g. [8]) that $\mu(A,C)$ defined by:

$$\mu(A,C) \triangleq \max\{\min\{ \|z\|_\infty \,|\, Az \leqq b,\ Cz = d\} \,|\, \|\binom{b}{d}\|_\infty \leqq 1,\ \binom{b}{d} \in F\} \tag{56}$$

is finite.

Theorem 5 [7]: Assume that the pair (A',C') satisfies the LI condition (see Assumption 2), and let $\mu = \mu(A',C')$. Let $\Delta A \in \mathbb{R}^{(q+1)\times n}$, $\Delta C \in \mathbb{R}^{m\times n}$ with $\delta \triangleq \|\binom{\Delta A}{\Delta C}\|_\infty$. If $\mu\delta < 1$, then the system (54), with $A = A' + \Delta A$, $C = C' + \Delta C$, has a solution for any right-hand side $\binom{b}{d}$. Moreover,

$$\mu(A,C) \leq \frac{\mu}{1-\mu\delta}. \quad \blacksquare \tag{57}$$

Proposition 2: Let $\mu:\mathbb{R}^{(q+1)\times n} \times \mathbb{R}^{m\times n} \to \mathbb{R}^+$ be defined by (56). Then μ is an upper-semicontinuous function. $\blacksquare$

Theorem 6: Let X be a compact subset of $\mathbb{R}^n$. Let $\bar{F}:X \to \mathbb{R}^{(q+1)\times n}$, $G: X \to \mathbb{R}^{m\times n}$ be continuous functions. Assume that for all $z \in X$, the pair $(\bar{F}(z), G(z))$ satisfies the LI condition. Then

$$\mu^* \triangleq \max_{z\in X} \mu[\bar{F}(z), G(z)] \tag{58}$$

exists and $\mu^* > 0$.

References

[1] Armijo, L. (1966) "Minimization of Functions Having Continuous Partial Derivatives," Pacific J. Math., 16, 1-3.

[2] Brent, R. P. (1973) "Some Efficient Algorithms for Solving Systems of Nonlinear Equations," SIAM J. Num. Anal., 10, 2, 327-344.

[2'] Huang, T. J. (1973) "Algorithms for Solving Systems of Equations and Inequalities with Applications in Nonlinear Programming," Computer Sciences Technical Report #191, Univ. of Wis.

[3] Kantorovich, L. V. and Akilov, G. P. (1964) Functional Analysis in Normed Spaces, McMillan, New York.

[4] Ortega, J. M. and Rheinboldt, W. C. (1970) Iterative Solution of Nonlinear Equations in Several Variables, Academic Press.

[5] Polak, E. (1971) Computational Methods in Optimization: A Unified Approach, Academic Press.

[6] Pshenichnyi, B. N. (1970) "Newton's Method for the Solution of Systems of Equalities and Inequalities," Math. Notes Acad. Sc. USSR, 8, 827-830.

[7] Robinson, S. M. (1971) "Extension of Newton's Method to Mixed Systems of Nonlinear Equations and Inequalities," Technical Summary Report No. 1161, Mathematics Research Center, Univ. of Wis.*

[8] Robinson, S. M. (1973) "Bounds for Error in the Solution Set of a Perturbed Linear Program," Lin. Alg. Appl., 6, 69-81.

[9] Shamanskii, V. (1967) "On a Modification of the Newton Method," Ukrain. Mat. J., 19, 133-138.

[10] Traub, J. (1964) Iterative Methods for the Solution of Equations, Prentice Hall, New Jersey.

*Editor's note: This report has been published, in revised form, in Numerische Mathematik 19 (1972), 341-347.

DISJUNCTIVE PROGRAMMING: CUTTING PLANES FROM LOGICAL CONDITIONS

by

Egon Balas

Abstract

Integer and many other nonconvex programming problems can be formulated as linear (or nonlinear) programs with logical conditions ("or," "and," negation of"). Since disjunction ("or") plays a crucial rule among these conditions, such problems are termed disjunctive programs. The family of all valid cutting planes for a disjunctive program is characterized. This family subsumes earlier cutting planes of the literature, and the procedure which generates the family offers clues to the strengthening of many of the earlier cuts. We also obtain entirely new cuts, which are computationally cheap and have other desirable properties, like coefficients of different signs. We discuss cutting planes for general mixed integer 0-1 programming, multiple choice problems, set partitioning, nonconvex quadratic programming, nonconvex separable programming.

1. Introduction

This paper summarizes some of the results of [5], where we discussed a class of cutting planes generated from disjunctions involving linear inequalities. The ideas underlying our developments were first presented in [4]. These ideas are rooted in our earlier work on intersection cuts [1], [2], [3], but several authors (Owen [22], [22a], Glover and Klingman [16], [16a], Zwart [24]) have obtained cutting planes from disjunctive constraints earlier. For contemporary developments closely related to ours see Glover [15] and Jeroslow [19].

Integer and many other nonconvex programming problems can be stated as disjunctive programs, a term by which we mean linear or nonlinear programs with logical constraints involving linear inequalities. Often this can be done in several ways. The logical constraints can of course be converted into linear inequalities involving 0-1 variables, but generating cuts directly from the logical conditions, without such a conversion, will be seen to have important advantages. The class of cutting planes that we obtain subsumes all of the earlier cuts proposed for integer and nonconvex programming, while improving some of them; and it includes a variety of new cuts with several desirable features.

Viewed somewhat differently, what we discuss is a new procedure for generating cutting planes, which has considerably greater power and flexibility than earlier approaches. This procedure can generate cuts whose coefficients have different signs, and which therefore are likely not to produce the well-known phenomenon of dual degeneracy that plagues all algorithms using sequences of cuts with coefficients of the same sign. Further, our approach can take full advantage of certain important and frequently occurring types of problem structure, and produce relatively cheap, yet powerful cuts for a variety of combinatorial and other nonconvex programming problems. These include generalized upper bounding, multiple choice problems, set partitioning, separable nonconvex programming, nonconvex quadratic programming, the linear complementarity problem, the generalized lattice point problem, etc. Last, but not least, this approach suggests new partitioning concepts for branch and bound [4], [6], which use the logical implications of the problem constraints to dichotomize the feasible set in ways which produce strong cuts, i.e., tight bounds, for the resulting subproblems.

2. Generating Cuts from Disjunctive Constraints

By logical conditions we mean in the present context statements about linear inequalities involving the operations "and" (conjunction), "or" (disjunction), "complement of" (negation), like for instance

$$"(A_1 \vee A_2) \wedge [(A_3 \wedge \bar{A}_4) \vee A_5]" ,$$

where $A_1, \ldots, A_5$ are linear inequalities, $\wedge$ and $\vee$ stand for conjunction and disjunction respectively, and $\bar{A}$ is the negation of A. A condition of the type "$A \Rightarrow B$" (implication) is known to be equivalent to the disjunction "$\bar{A} \vee B$." Since the operations $\wedge$ and $\vee$ are distributive with respect to each other, any logical condition of the above type can be restated as a disjunction between sets of simultaneous inequalities, i.e., brought to the <u>disjunctive normal form</u> "at least one of several sets of inequalities must hold." The disjunctive normal form of the above example is

$$"A_1A_3\bar{A}_4 \vee A_1A_5 \vee A_2A_3\bar{A}_4 \vee A_2A_5" ,$$

where the symbol $\wedge$ of conjunction has been replaced by juxtaposition.

The operations of conjunction and negation applied to linear inequalities give rise to (convex) polyhedral sets and hence leave the problem of optimizing a linear form subject to such constraints within the realm of linear programming. It is the disjunctions that introduce nonconvexities; that is why we view them as the crucial element and call the problems in this class disjunctive programs.

Let R^n_+ be the nonnegative orthant of R^n, let T be an arbitrary subset of R^n_+, and consider the constriant set

$$F = \{x\varepsilon T | \bigvee_{h\varepsilon H} (\sum_{j\varepsilon N} a^h_{ij}x_j \geq a^h_{i0}, i\varepsilon Q_h)\}$$

where Q_h, $h \varepsilon H$, are finite index sets, H is a (not necessarily finite) index set, and the coefficients a^h_{ij}, $i \varepsilon Q_h$, $h \varepsilon H$, $j \varepsilon N \cup \{0\}$, are real numbers. Since T is not required to be linear, F is the constraint set of a genreal (nonlinear) disjunctive program.

Let H^* be the set of those $h \varepsilon H$ such that

$$\{x\varepsilon T | \sum_{j\varepsilon N} a^h_{ij}x_j \geq a^h_{i0}, i\varepsilon Q_h\} \neq \emptyset .$$

Theorem 1. Every $x \in T$ which satisfies at least one of the systems

$$(1)_h \qquad \left\{ \sum_{j \in N} a^h_{ij} x_j \geq a^h_{i0}, \ i \in Q_h \right\}, \ h \in H ,$$

also satisfies the inequality

$$(2) \qquad \sum_{j \in N} \alpha_j x_j \geq \alpha_0$$

for every finite α_0 and α_j, $j \in N$, such that

$$(3) \qquad \alpha_j \geq \sup_{h \in H^*} \sum_{i \in Q_h} \theta^h_i a^h_{ij} , \ j \in N ,$$

and

$$(4) \qquad \alpha_0 \leq \inf_{h \in H^*} \sum_{i \in Q_h} \theta^h_i a^h_{i0}$$

for some $\theta^h_i \geq 0$, $i \in Q_h$, $h \in H^*$.

Proof. Suppose $\hat{x} \in T$ satisfies $(1)_k$ for some $k \in H$. By adding the inequalities of $(1)_k$ after multiplying the i-th inequality by some $\theta^k_i \geq 0$, $i \in Q_k$, we obtain

$$\begin{aligned} \alpha_0 &\leq \inf_{h \in H^*} \sum_{i \in Q_h} \theta^h_i a_{i0} \\ &\leq \sum_{i \in Q_k} \theta^k_i a_{i0} \qquad \text{(since } k \in H^*\text{)} \\ &\leq \sum_{j \in N} \left(\sum_{i \in Q_k} \theta^k_i a^k_{ij} \right) \hat{x}_j \end{aligned}$$

$$\leq \sum_{j\varepsilon N} (\sup_{h\varepsilon H^*} \sum_{i\varepsilon Q_h} \theta_i^h a_{ij}^h)\hat{x}_j \quad (\text{since } k\varepsilon H^*, \text{ and } \hat{x} \varepsilon T \Rightarrow \hat{x} \geq 0)$$

$$\leq \sum_{j\varepsilon N} \alpha_j \hat{x}_j \ . \qquad \underline{\text{Q.E.D.}}$$

Remark 1. Theorem 1 remains true if the i-th inequality of $(1)_h$, for some $h\varepsilon H$, is replaced by an equation, and, if $h\varepsilon H^*$, the nonnegativity requirement on the variable θ_i^h is removed.

The importance of the class of cuts defined by Theorem 1 lies in its generality. By an appropriate choice of the multipliers θ_i^h, most of the earlier cuts proposed in the literature can easily be recovered, many of them in a strengthened version. Furthermore, the following (partial) converse of Theorem 1, due to R. Jeroslow, shows that if $T = R_+^n$ (i.e., we are dealing with a linear disjunctive program), then the class of cutting planes defined by (2) contains among its members the strongest possible cuts implied by the disjunction, i.e., the facets of the convex hull of all $x \varepsilon R_+^n$ satisfying at least one of the systems $(1)_h$, $h \varepsilon H$. A characterization of these facets is given in [7].

(Partial Converse of Theorem 1 (Jeroslow). Let $T = R_+^n$. Then every valid cut, i.e., every inequality that holds for all $x \varepsilon R_+^n$ satisfying at least one of the systems $(1)_h$, $h \varepsilon H$, is of the form (2), with α_0 and α_j, $j \varepsilon J$, defined by (3) and (4).

Proof. An inequality $\sum_{j\varepsilon N} \alpha_j x_j \geq \alpha_0$ which holds for all $x \varepsilon R^n_+$ satisfying at least one of the systems $(1)_h$, $h \varepsilon H$, is a consequence of each system $(1)_h$, $h \varepsilon H^*$, amended with the condition $x \geq 0$. Since each of these systems is consistent, according to a classical result on linear inequalities (see, for instance, Theorem 22.2 in Rockafellar [23]), for each $h \varepsilon H^*$ there exists a set of multipliers $\theta^h_i \geq 0$, $i \varepsilon Q_h \cup \{j\}$, such that

$$\alpha_j = \sum_{i\varepsilon Q_h} \theta^h_i a^h_{ij} + \theta^h_j , \quad j \varepsilon J, \ h \varepsilon H^* ,$$

and

$$\alpha_0 \leq \sum_{i\varepsilon Q_h} \theta^h_i a^h_{i0} , \quad h \varepsilon H^* .$$

Since $\theta^h_j \geq 0$, $j \varepsilon J$, $h \varepsilon H^*$, (3) and (4) follows. Q.E.D.

Remark 1 is of course as valid for the converse of Theorem 1 as it is for the theorem itself.

We now examine some immediate consequences of Theorem 1. Suppose first that each system $(1)_h$, $h \varepsilon H$ (where again H need not be finite), consists of a single inequality with coefficients a_{hj}, $j \varepsilon N \cup \{0\}$, such that $a_{h0} > 0$, $\forall\, h \varepsilon H$, and let H^* be defined as before. Then we have

Corollary 1.1. Every $x \varepsilon T$ which satisfies at least one of the inequalities

$$\sum_{j\varepsilon N} a_h x_j \geq a_{h0} , \quad h \; \varepsilon \; H$$

also satisfies the inequality

$$\sum_{j\varepsilon N} \left(\sup_{h\varepsilon H^*} \frac{a_{hj}}{a_{h0}} \right) x_j \geq 1 . \tag{5}$$

Proof. Follows from Theorem 1 when $|Q_h| = 1, \forall \; h \; \varepsilon \; H$. Q.E.D.

When T is linear, and $H^* = H$, the cut (5) is an improved version of the usual intersection (or convexity) cut obtained by intersecting the n half-lines $\xi_j = \{x|x = e_j\lambda_j, \lambda_j \geq 0\}$, $j \; \varepsilon \; N$ (where e_j is the j-th unit vector in R^n), with the boundary of the closed convex set

$$S = \{x\varepsilon R^n | \sum_{j\varepsilon N} a_{hj}x_j \leq a_{h0}, h\varepsilon H\}.$$

For those $j \; \varepsilon \; N$ such that $\xi_j \cap \text{bd } S \neq \emptyset$, the coefficient of (5) is positive, and the same as for the usual intersection cut. For those $j \; \varepsilon \; N$ such that $\xi_j \cap \text{bd } S = \emptyset$, the usual intersection cut has a zero coefficient, whereas (5) may have a negative one, corresponding to the intersection of $-\xi_j$ with the last supporting hyperplane of S (see [5] for details).

Needless to say, the cut (5) remains valid if H^* is replaced by H , and whenever the two sets are different, this replacement is a weakening.

For the case when T is linear and S is polyhedral (i.e., H is finite), the version of (5) in which H^* is replaced by H is the cut found by Owen [22] (see also Glover [14]).

Another important case we wish to examine is the one when $|Q_h| = k$, $\forall\, h \in H$, for some integer k. This case arises from conditions of the type "at least k inequalities of a set Q must hold." It is then possible to choose the weights θ_i^h in ways which produce cuts of a particularly simple form. Let again $T \subseteq R_+^n$, and let Q^* be the set of those $i \in Q$ such that

$$\{x \in T \mid \sum_{j \in N} a_{ij} x_j \geq a_{i0}\} \neq \emptyset .$$

<u>Corollary 1.2</u>. Every $x \in T$ which satisfies (at least) k inequalities of the set

$$\sum_{j \in N} a_{ij} x_j \geq a_{i0}, \quad i \in Q ,$$

where $a_{i0} > 0$, $\forall\, i \in Q$, and $1 \leq k \leq |Q|$, also satisfies

$$\sum_{j \in N} \beta_j x_j \geq 1 \tag{6}$$

and

$$\sum_{j \in N} \gamma_j x_j \geq 1 , \tag{7}$$

where

$$\beta_j = \max_{K \subseteq Q^* \,|\, |K| = k} \left\{ \frac{1}{k} \sum_{i \in K} \frac{a_{ij}}{a_{i0}} \right\}$$

and

$$\gamma_j = \max_{i\varepsilon Q'} \frac{a_{ij}}{a_{i0}}$$

Q' being any set (the same for each $j\varepsilon N$) obtained from Q^* by deleting k-1 of its elements.

Proof. Let Q_h^*, $h \varepsilon H$, be all subsets of Q^* of cardinality k. Then restating the condition of the Corollary in disjunctive normal form and applying Theorem 1 we obtain an inequality of the form (2), with

$$\alpha_j \geq \sup_{h\varepsilon H} \sum_{i\varepsilon Q_h^*} \theta_i^h a_{ij}, \; j \varepsilon J$$

and

$$\alpha_0 \leq \inf_{h\varepsilon H} \sum_{i\varepsilon Q_h^*} \theta_i^h a_{i0} ,$$

where $\theta_i^h \geq 0$, $i \varepsilon Q_h^*$, and $|Q_h^*| = k$, $\forall h \varepsilon H$.

Setting $\theta_i^h = \frac{1}{ka_{i0}}$, $\forall h \varepsilon H$, yields

$$\sum_{i\varepsilon Q_h^*} \theta_i^h a_{i0} = 1, \; \forall h \varepsilon H ,$$

and the coefficients β_j of (6).

On the other hand, deleting any k-1 inequalities of the set indexed by Q^* and caling the remaining set Q', every $x \varepsilon T$ which satisfies at least k inequalities among those indexed by Q^*, also satisfies at least one inequality of those indexed by Q'. Applying Corollary 1.1 then yields the cut (7). Q.E.D.

Note that both cuts (6) and (7) are remarkably easy to calculate, especially in their (weaker) form where Q^* is replaced by Q . For (7), it suffices to remove any $k - 1$ of the inequalities indexed by Q , and take for γ_j the largest of the coefficients $\frac{a_{ij}}{a_{i0}}$ in the remaining set. For (6), β_j is simply the arithmetic mean of the k largest coefficients $\frac{a_{ij}}{a_{i0}}$ among all $i \varepsilon Q$.

If $k = 1$, then (6) and (7) are both the same as the cut (4) of Corollary 1.1. In general, for small k (7) is likely to be stronger than (6), since by deleting the rows which contain the largest number of column-maxima one can usually do better than by averaging coefficients. On the contrary, for k close to $|Q|$, (6) is likely to be stronger than (7), since taking the arithmetic mean of k numbers $\frac{a_{ij}}{a_{i0}}$ of arbitrary sign is likely to produce coefficients β_j small in absolute value (which, with a constant right hand side of 1, means a strong cut), whose signs are about equally likely to be + or - .

In the rest of this paper we apply the results of section 2 to various nonconvex programming problems formulated as linear programs with disjunctive constraints.

3. Mixed Integer 0-1 Programs

We start by adapting Theorem 1 and its converse to the case of a general mixed integer program with 0-1 variables.

Theorem 2. Let S be the set of points $(x,y) \in R^n \times R^q$ defined by

$$(8) \qquad x_i = a_{i0} + \sum_{j\in J_1} a_{ij}(-x_j) + \sum_{j\in J_2} a_{ij}(-y_j), i\in I_1$$

$$(9) \qquad y_i = a_{i0} + \sum_{j\in J_1} a_{ij}(-x_j) + \sum_{j\in J_2} a_{ij}(-y_j), i\in I_2$$

$$(10) \qquad x_i = 0 \text{ or } 1, \; i \in N = I_1 \cup J_1$$

$$(11) \qquad y_i \geq 0, \; i \in Q = I_2 \cup J_2 ,$$

where $n = |N|$, $q = |Q|$, and let $I = I_1 \cup I_2$, $J = J_1 \cup J_2$. Further, let

$$X = \{x\in R^n | (x,y)\in S \text{ for some } y\in R^q\}.$$

Then the inequality

$$(12) \qquad \sum_{j\in J_1} \alpha_j x_j + \sum_{j\in J_2} \alpha_j y_j \geq \alpha_0$$

where $-\infty < \alpha_j < \infty$, $j \in J \cup \{0\}$, is satisfied by all $(x,y) \in S$, if and only if

$$(13) \qquad \alpha_j \geq \sup_{x\in X} \{\theta_j(x) - \sum_{i\in I} a_{ij}\theta_i(x)\} , \; j \in J$$

and

$$(14) \qquad \alpha_0 \le \inf_{x \in X} \left\{ \sum_{i \in J_1} x_i \theta_i(x) + \sum_{i \in I_1} (x_i - a_{i0}) \theta_i(x) + \sum_{i \in I_2} (-a_{i0}) \theta_i(x) \right\}$$

for some function $\theta(x): X \to R^{n+q}$, such that

$$-\infty < \theta_i(x) < \infty, \; i \in N, \; 0 \le \theta_i(x) < \infty, i \in Q.$$

Proof. The conditions defining S can be stated in disjunctive normal form by requiring every $(x,y) \in S$ to satisfy at least one of the sets of constraints consisting of (8), (9), (11), and $x = x^h \in X$, for some $h \in H$, where H indexes the elements of X. Applying Theorem 1 (with Remark 1) we find that (12) is satisfied by all $(x,y) \in S$ if

$$\alpha_j \ge \sup_{h \in H} \left\{ \theta_j^h + \sum_{i \in I} \theta_i^h (-a_{ij}) \right\}, \; j \in J$$

and

$$\alpha_0 \le \inf_{h \in H} \left\{ \sum_{i \in J_1} \theta_i^h x_i^h + \sum_{i \in I_1} \theta_i^h (x_i^h - a_{i0}) + \sum_{i \in I_2} \theta_i^h (-a_{i0}) \right\}$$

for some θ_i^h, $i \in N \cup Q$, $h \in H$, such that

$$-\infty < \theta_i^h < \infty, \; i \in N, \; 0 \le \theta_i^h < \infty, \; i \in Q, \; \forall h \in H.$$

Then denoting $\theta_i^h = \theta_i(x^h)$, $\forall\, x^h \in X$, $i \in N \cup Q$, we get (13) and (14), and we obtain the "if" part of Theorem 2. Also, applying the converse of Theorem 1 yields the "only if" part. Q.E.D.

Theorem 2 states the general form of any valid inequality for a mixed-integer 0-1 program. For the inequality (12) to cut off the point $(\bar{x},\bar{y})$, defined by $\bar{x}_i = a_{i0}$, $i \in I_1$, $\bar{x}_i = 0$, $i \in J_1$, $\bar{y}_i = a_{i0}$, $i \in I_2$, $\bar{y}_i = 0$, $j \in J_2$, it is necessary and sufficient, in addition to the conditions of the Theorem, that $\alpha_0 > 0$. On the other hand, all valid cuts, including those facets of the convex hull of S which cut off $(\bar{x},\bar{y})$, are of the form (12). Of course, there are many ways of choosing the parameters $\theta_i(x)$, and the theorem leaves open the question as to which particular sets of parameters yield facets. This question, which is important from a theoretical point of view, is settled in [7]. However, the fact that determining the smallest possible coefficients α_j involves a maximization over X suggests that finding the facets of the integer hull for a general mixed-integer 0-1 program is a difficult task and there is probably no polynomially bounded procedure for doing it. On the other hand, the point should not be missed that the maximization problems involved define lower bounds on the coefficients α_j, and therefore one can replace X with a set of weaker (relaxed) constraints and still obtain valid and strong (though not the strongest possible) cuts via Theorem 2. One possible relaxation of X is to waive the integrality requirement, (i.e., solve linear instead of integer programs, in a vein similar to [2]); another one is to remove some or all of the inequalities defining X. Further, the maximization procedure itself can be replaced by any heuristic which yields an upper bound for the maximand.

In this context, it is interesting and in most cases not too difficult to identify those values of the parameters which yield various cuts proposed earlier in the literature, and such identification is useful because, by putting an old cut in the new context of the present approach, i.e., by establishing which special case of the inequality (12) it is, one also finds directions for possible improvements upon the given cut. Thus, in [5] we discuss ways of recovering and improving Gomory's mixed integer cuts [17], Burdet's diamond cuts [11] which are convex combinations of the latter, as well as the polar cuts of [2], [3] and [8].

One important feature of the improvement that is obtained over cuts proposed earlier is that one can obtain coefficients with different signs. It is well known that dual cutting plane algorithms tend to break down because of massive dual degeneracy. This is due to the fact that the cutting planes (inequalties ≥) traditionally used in such algorithms have all their coefficients nonnegative, and most of them positive (i.e., when restated in simplex tableau format, negative). Thus the first pivot after adding each new cut decreases most if not all the coefficients of the zero row, and never increases any of them (with the possible exception of the pivot column). This gradually leads to more and more coefficients a_{0j} being equal to or close to zero. Geometrically, a cut whose coefficients are all positive, corresponds to the case where each of the edges of the cone associated with the current LP-solution intersects the cutting plane. The repeated use of cutting planes intersected by all or most edges of the linear programming cone, leads to a "flattening" of the feasible set in the region of the cuts, and the successive cutting planes tend to be more and more parallel to the objective function hyperplane. Fig. 1 illustrates in 2 dimensions how successive cuts produce more and more obtuse angles at the current linear programming optimum.

On the other hand, cutting planes with both positive and negative coefficients produce pivots which decrease some, but increase other coefficients of the zero row. The possibility of systematically generating such cuts is of paramount importance, since it makes it possible to generate sequences of cutting planes accompanied by dual pivots without necessarily creating a tendency towards degeneracy. Geometricaly, such cuts correspond to the situation illustrated in Fig. 2, where some of the edges of the linear programming cone have to be extended in the negative direction to intersect the cutting plane. Here the successive angles produced, far from becoming more and more obtuse, tend to become more acute.

There are many new cuts that can be obtained by our approach, at a relatively low computational cost. A very promising class of cutting planes can be derived by restating as a logical condition some canoncial inequality (see [9]) of the form

$$\sum_{j \in Q} x_j \leq |Q| - k ,$$

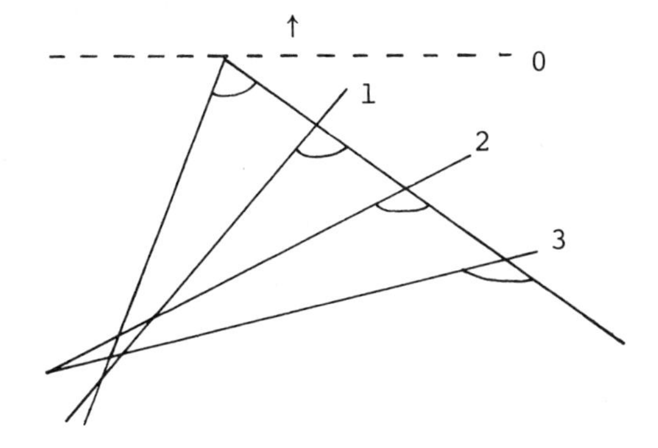

0 objective function
1 first cut
2 second cut
3 third cut

Figure 1

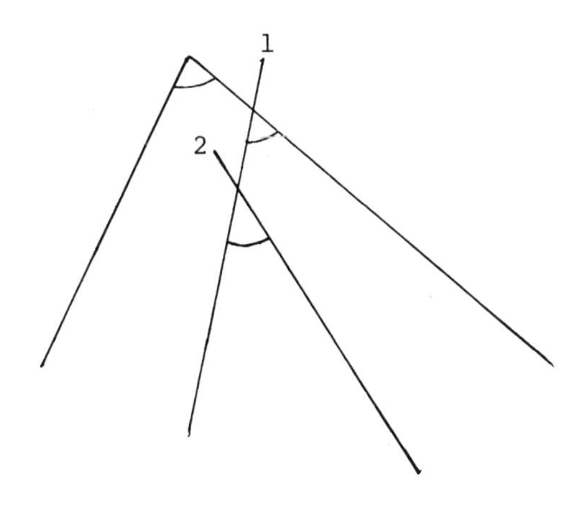

1 first cut
2 second cut

Figure 2

implied by the problem constraints (here $Q \subseteq N$ and $1 \leq k \leq |Q| - 1$). This can be done in several ways, like

"$x_j \leq 0$ for at least k indices $j \in Q$,"

or, choosing some subset $Q' \subset Q$ such that $|Q'| = |Q| - k$,

"either $x_j \geq 1, \forall j \in Q'$, and $x_j \leq 0, \forall j \in Q - Q'$,

or $x_j \leq 0$ for at least one $j \in Q'$."

The derivation and properties of such cuts are discussed in [5]. We mention here that conditions of the latter type offer excellent opportunities for combining cutting planes with branch and bound in promising new ways. For instance, the last disjunction may be split into two, i.e., one may partition the problem into two subproblems, the first of which satisfies "$x_j = 1, \forall j \in Q', x_j = 0, \forall j \in Q - Q'$," i.e., has $|Q|$ variables fixed, whereas the second one satisfies the inequality (cut) generated from "$x_j \leq 0$ for at least one $j \in Q'$." With an appropriate choice of Q', the latter cut can have very attractive properties (like yielding a very high penalty, or having coefficients of different signs, etc.). This and related issues are discussed in [6].

Next we turn to specially structured 0-1 programs and other nonconvex problems for which the present approach yields strong and easily obtainable cuts.

4. Set Partitioning and Multiple Choice Constraints

The set partitioning, or equality constrained set covering problem, is a good example for illustrating the power of the approach discussed in this paper. More generally, the results of this section apply to any mixed integer program with at least one constraint of the form

$$\sum_{i \in Q} x_i = 1$$

where the variables x_i , $i \in Q$, are integer-constrained. This, of course, encompasses the so called multiple choice constraints, which can always be brought to the above form by inserting a slack variable.

Theorem 3. Given a mixed-integer program whose constraints include or imply

$$\text{(15)} \qquad \sum_{i \in Q} x_i = 1, \; x_i \geq 0, \text{ integer}, \; i \in Q ,$$

and a basic feasible solution to the associated linear program, let

$$x_i = a_{i0} + \sum_{j \in J} a_{ij}(-t_j)$$

be the expression of x_i , $i \in Q$, in terms of the nonbasic variables, and let $Q_* = \{i \in Q \,|\, 0 < a_{i0} < 1\} \neq \emptyset$.

Then every feasible integer solution satisfies the inequality

$$\text{(16)} \qquad \sum_{j \in J} \alpha_j t_j \geq 1$$

for each of the following three definitions of its coefficients:

$$(17)\quad \alpha_j = \begin{cases} \dfrac{1}{|Q_*|} \displaystyle\sum_{i\varepsilon Q_*} \dfrac{a_{ij}}{a_{i0}} & , \text{ if } j\varepsilon J\cap Q, \text{ or } a_{ij}\geq 0,\ \forall\ i\varepsilon Q \\ \dfrac{1}{|Q_*|}\left[\displaystyle\sum_{i\varepsilon Q_*} \dfrac{a_{ij}}{a_{i0}} - \min_{i\varepsilon Q_*} \dfrac{a_{ij}}{a_{i0}(1-a_{i0})}\right] & , \text{ if } j\varepsilon J\sim Q \text{ and } a_{ij}<0 \text{ for some } i\varepsilon Q; \end{cases}$$

$$(18)\quad \alpha_j = \max_{i\varepsilon Q} \frac{1}{1-a_{i0}}\left(\sigma_i \sum_{h\varepsilon Q} a_{hj} - a_{ij}\right),\ \forall\ j\varepsilon J,$$

where the σ_i are any numbers such that $0 \leq \sigma_i \leq 1$ (identical for all $j\varepsilon J$);

$$(19)\quad \alpha_j = \max_{i\varepsilon Q_{**}} \frac{a_{ij}}{a_{i0}}$$

where Q_{**} is any subset of Q_* such that $|Q_{**}| = 2$.

<u>Proof</u>. (i) The condition (15) is equivalent to the disjunction

$$\begin{Bmatrix} x_h \leq 0,\ \forall\ h\varepsilon Q - \{1\} \\ x_1 \geq 1 \end{Bmatrix} \vee \cdots \vee \begin{Bmatrix} x_h \leq 0,\ \forall\ h\varepsilon Q - \{q\} \\ x_q \geq 1 \end{Bmatrix}$$

where $\{1,\ldots,q\} = Q$.

Rewriting this in terms of the variables t_j, $j\ \varepsilon\ J$, and applying Theorem 1, yields the inequality (16) with

$$(20)\quad \alpha_j \geq \sup_{i\varepsilon Q} \left\{\sum_{h\varepsilon Q-\{i\}} \theta_h^i a_{hj} - \theta_i^i a_{ij}\right\},\ j\ \varepsilon\ J$$

and

$$\alpha_0 \leq \sup_{i\varepsilon Q} \left\{ \sum_{h\varepsilon Q-\{i\}} \theta_h^i a_{h0} + \theta_i^i (1-a_{i0}) \right\} ,$$

where $\theta_h^i \geq 0, \forall\ h \ \varepsilon\ Q$.

To obtain (17), we set

$$\theta_h^i = \begin{cases} a_{h0}^{-1} |Q_*|^{-1} & h\varepsilon Q_* \sim \{i\} \\ (1-a_{h0})^{-1} |Q_*|^{-1} & h\varepsilon Q_* \cap \{i\} \\ L & h\varepsilon \bar{Q}_* \cap J \sim \{i\} \\ 0 & h\varepsilon \bar{Q}_* \cap (I\cup\{i\}) \end{cases}$$

where Q_* and J are as above, I is the basic index set, $\bar{Q}_* = Q - Q_*$, and L is a large number. By definition, $\theta_h^i \geq 0, \forall\ h,i$. Further, since the current solution is feasible, $\sum_{h\varepsilon Q} a_{h0} = 1$ and thus $a_{h0} = 0, \forall\ h\varepsilon \bar{Q}_*$. Therefore

$$\sum_{h\varepsilon Q-\{i\}} \theta_h^i a_{h0} + \theta_i^i (1-a_{i0}) = 1, \forall\ i\varepsilon Q, \text{ i.e., } \alpha_0 = 1 .$$

On the other hand, by substituting for θ_h^i in (20), and taking the supremum over all $i\varepsilon Q$, one obtains the expression (17) for α_j, $j\varepsilon J$ (this is not immediate; details are given in [5]).

(ii) To obtain (18), we set

$$\theta_h^i = \begin{cases} \sigma_i (1-a_{i0})^{-1} & h \neq i \\ (1-\sigma_i)(1-a_{i0})^{-1} & h = i \end{cases} \qquad \forall\ i\varepsilon Q$$

where $0 \leq \sigma_i \leq 1$. This satisfies the requirements on the multipliers θ_h^i, and produces the desired result.

(iii) Finally, (19) is simply a special case of the coefficient γ_j for the cut (7) of Corollary 12, after replacing the set Q^* of that Corollary by Q (which is justified as a weakening). Here $k = q - 1$ and therefore the maximum has to be taken over all subsets of Q of cardinality $q - (k-1) = 2$. Q.E.D.

Each of the three cuts given in the Theorem is likely to have a number of negative coefficients. The cut defined by (19) is the easiest to compute, and the two rows indexed by Q_{**} should be selected so as to have the largest possible number of negative entries in overlapping columns, so as to give rise to a cut with as many negative coefficients as possible.

The other two cuts are likely to be stronger than the one defined by (19), and still relatively cheap to compute. The potential strength of the cut defined by (17) can be particularly well assessed, since its coefficients are close to the arithmetic means of the coefficients $\frac{a_{ij}}{a_{i0}}$, $i \in Q_*$, of the given column. Since the coefficients may be of any sign, the likelihood of obtaining a cut with small coefficients (for fixed right hand side) increases with $|Q_*|$. This cut [the one defined by (17)] compares favorably with the cut from the outer polar of the truncated cube defined by $\sum_{i \in Q} x_i \leq 1$,

$0 \le x_i \le 1$, $i \in Q$, which in turn was shown (see [10]) to dominate the cuts of [13]. Besides, the present cut is considerably easier to compute than either of the other two.

As for the cut defined by (18), it can be shown (see [5]) to dominate every convex combination of the two cuts derived from the disjunctions "$x_h \le 0$, $h \in Q - \{i\}$, for at least one $i \in Q$," and "$x_i \ge 1$ for at least one $i \in Q$," respectively.

Finally, we mention that if one sets $\sigma_i = 1 - a_{i0}$, $\forall\, i \in Q$, then (18) becomes

$$(18') \qquad \alpha_j = \sum_{h \in Q} a_{hj} - \min_{i \in Q} \frac{a_{ij}}{1-a_{i0}} ,$$

which is particularly easy to compute.

5. Nonconvex Separable Programming

It is well known that nonlinear programs involving only separable functions, i.e., functions of the form $f(u) = \sum_{j=1}^{m} f_j(u_j)$, can be approximated (with whatever degree of precision seems desirable) by piecewise linearization and reduction to a linear program with an additional disjunctive constraint. This can be done in (at least) two different ways, known in the literature by the odd names of "λ-form" and "δ-form" of approximation.

The "λ-form" of approximation results in a linear program whose variables $x_i \ge 0$, $i \in Q = \{1,\ldots,q\}$, $Q \subseteq N$, are subject (besides other linear constraints) to the equation $\sum_{i \in Q} x_i = 1$ and the following special constraint (which we will call the "λ-condition"): "there can be at most two positive x_i, $i \in Q$, and if there are two, their indices differ by 1."

The "δ-form" of approximation leads to a linear program whose variables $x_i \geq 0$, $i \in Q, = \{1,\ldots,q\}$, $Q \subseteq N$, are subject (besides other linear constraints) to upper bounding constraints $x_i \leq d_i$, $i \in Q$ and the following special constraint (which we will call the "δ-condition"): "$x_i > 0 \Rightarrow x_j = d_j$, $j = 1,\ldots,i-1$, for all $i \in \{2,\ldots,q\}$."

In each of the above two cases, the special constraint can be put in disjunctive normal form and used to generate cutting planes.

We first deal with the "λ-form."

Theorem 4. Given a linear program whose constraints include or imply the equation

$$(21) \quad \sum_{i\in Q} x_i = 1 ,$$

where $Q = \{1,\ldots,q\}$, let

$$x_i = a_0 + \sum_{j\in J} a_{ij}(-t_j)$$

be the i-th row of the simplex tableau associated with a basic feasible (i.e., $a_{i0} > 0$, $\forall$ i) solution which violates the λ-condition. Then every solution that satisfies the λ-condition also satisfies the inequality

$$(22) \quad \sum_{j\in J} \alpha_j t_j \geq 1$$

where

$$(23)\quad \alpha_j = \max_{i\varepsilon Q-\{q\}} (1-a_{i0}-a_{i+1,0})^{-1}(\sigma_i \sum_{h\varepsilon Q} a_{hj}-a_{ij}-a_{i+1,j})$$

for any σ_i, $i \varepsilon Q - \{q\}$, such that $0 \le \sigma_i \le 1$ (identical for all $j \varepsilon J$).

Proof. The λ-condition together with (21) can be stated as

$$\left\{\begin{matrix} x_h \le 0, \ h = 3,\dots,q \\ x_1 + x_2 \ge 1 \end{matrix}\right\} \vee \dots \vee \left\{\begin{matrix} x_h \le 0, \ h = 1,\dots,q-2 \\ x_{q-1} + x_q \ge 1 \end{matrix}\right\}.$$

Restating this disjunction in terms of the nonbasic variables and applying Theorem 1 yields the cut (22), with

$$\alpha_j \ge \sup_{i\varepsilon\{1,\dots,q-1\}} \left\{ \sum_{h\varepsilon Q-\{i,i+1\}} \theta^i_h a_{hj} - \theta^i_i (a_{ij}+a_{i+1,j}) \right\}, \quad j\varepsilon J,$$

and

$$\alpha_0 \le \inf_{i\varepsilon\{1,\dots,q-1\}} \left\{ \sum_{h\varepsilon Q-\{i,i+1\}} \theta^i_h a_{h0} + \theta^i_i (1-a_{i0}-a_{i+1,0}) \right\},$$

where $\theta^i_h \ge 0$, $h \varepsilon Q - \{i+1\}$.

To obtain (23), we then set

$$\theta^i_h = \begin{cases} \sigma_i (1-a_{i0}-a_{i+1,0})^{-1} & h \ne i, \ i+1 \\ (1-\sigma_i)(1-a_{i0}-a_{i+1,0})^{-1} & h = i \end{cases}$$

where $0 \leq \sigma_i \leq 1$, $\forall\, i \in Q - \{q\}$. Since $1 - a_{i0} - a_{i+1,0} > 0$, $\forall\, i \in Q - \{q\}$ (from the fact that the current solution violates the λ-condition), this satisfies the requirements of Theorem 1 on the multipliers θ_h^i, $\forall$ h, i, and yields (23). <u>Q.E.D.</u>

Again, the cut (22) can be shown (see [5]) to dominate every convex combination of the two cuts obtained from the disjunctions "$x_h \leq 0$, $\forall\, h \in Q - \{i,i+1\}$ for at least one $i \in Q - \{q\}$," and "$x_i + x_{i+1} \geq 1$ for at least one $i \in Q - \{q\}$," respectively.

Notice that, if one sets $\sigma_i = 1 - a_{i0} - a_{i+1,0}$, $\forall\, i \in Q - \{q\}$, then

$$(23') \qquad \alpha_j = \sum_{h\in Q} a_{hj} - \min_{i\in Q-\{q\}} \frac{a_{ij}+a_{i+1,j}}{1-a_{i0}-a_{i+1,0}} .$$

Next we address ourselves to the "δ-form" of approximation.

<u>Theorem 5</u>. Given a linear program whose constraints include or imply the inequalities

$$(24) \qquad x_i \leq d_i, \; i \in Q ,$$

where $Q = \{1,\ldots,q\}$, let

$$x_i = a_{i0} + \sum_{j\in J} a_{ij}(-t_j)$$

be the i-th row of the simplex tableau associated with a basic feasible (i.e., $a_{i0} \geq 0$, $\forall$ i) solution which violates the δ-condition. Then every feasible solution that satisfies the δ-condition also satisfies the inequality

$$\sum_{j\varepsilon J} \alpha_j t_j \geq 1 , \qquad (25)$$

for each of the following two definitions of its coefficients:

$$\alpha_j = \max_{i\varepsilon Q} \frac{\sum_{h=1}^{i-1} (-a_{hj}) + \sum_{h=i+1}^{q} a_{hj}}{\sum_{h=1}^{i-1} (d_h - a_{ho}) + \sum_{h=i+1}^{q} a_{ho}} \qquad (26)$$

$$\alpha_j = \max_{i\varepsilon Q} \frac{1}{|Q_*^i|} \left(\sum_{h\varepsilon Q_*^{i1}} \frac{-a_{hj}}{d_h - a_{ho}} + \sum_{h\varepsilon Q_*^{i2}} \frac{a_{hj}}{a_{ho}} \right) \qquad (27)$$

where

$$Q_*^{i1} = \{1,\ldots,i-1\} \cap \{h\varepsilon Q \mid 0 \leq a_{ho} < d_h\},$$

$$Q_*^{i2} = \{i+1,\ldots,q\} \cap \{h\varepsilon Q \mid 0 < a_{ho} \leq d_h\},$$

and $Q_*^i = Q_*^{i1} \cup Q_*^{i2}$.

<u>Proof</u>. The "δ-condition" together with (24) can be stated as the disjunction "at least one of the systems

$$(28)_i \qquad \left\{ \begin{array}{ll} x_h \geq d_h & h = 1,\ldots,i-1 \\ x_h \leq 0 & h = i+1,\ldots,q \end{array} \right\}$$

$i = 1,\ldots,q$ (where $x_h \geq d_h$ is vacuous when $i = 1$, and $x_h \leq 0$ is vacuous when $i = q$), must hold." Restating $(28)_i$ in terms of the nonbasic variables and applying Theorem 1, we obtain the cut (25) with

$$(29) \qquad \alpha_j \geq \sup_{i\epsilon Q} \left\{ \sum_{h=1}^{i-1} \theta_h^i(-a_{hj}) + \sum_{h=i+1}^{q} \theta_h^i a_{hj} \right\},$$

and

$$(30) \qquad \alpha_o \leq \inf_{i\epsilon Q} \left\{ \sum_{h=1}^{i-1} \theta_h^i(d_h - a_{ho}) + \sum_{h=i+1}^{q} \theta_h^i a_{ho} \right\},$$

where $\theta_h^i \geq 0, \forall\ h\ \epsilon\ Q - \{i\}$.

The multipliers

$$\theta_h^i = \theta^i = \left[\sum_{h=1}^{i-1} (d_h - a_{ho}) + \sum_{h=i+1}^{q} a_{ho} \right]^{-1},$$

$$h\ \epsilon\ Q - \{i\},\ i\ \epsilon\ Q\ ,$$

satisfy the above requirements and, when substituted into (30), yield equality for $\alpha_o = 1$. Also, when substituted in (29), they yield the coefficients α_j defined by (26).

On the other hand, the multipliers

$$\theta_h^i = \begin{cases} (d_h - a_{ho})^{-1} |Q_*^i|^{-1} & h\epsilon Q_*^{i1} \\ a_{ho}^{-1} |Q_*^i|^{-1} & h\epsilon Q_*^{i2} \\ 0 & h\epsilon Q - Q_*^i,\ i\epsilon Q\ , \end{cases}$$

which also satisfy the requirements, yield (27).

Q.E.D.

6. Nonconvex Quadratic Programming and the Linear Complementarity Problem

Consider the general linearly constrained quadratic programming problem

$$(P) \qquad \max\{cx + \frac{1}{2} xCx | Ax \leq b,\ x \geq 0\}$$

where C is a symmetric $r \times r$ matrix, and A an arbitrary $m \times r$ matrix. Notice that C is allowed to be indefinite.

According to basic nonlinear programming theory (see, for instance, Theorem 7.3.7 of [21]), if $\bar{x}$ is a local maximum of (P), then there exists $\bar{y}$, $\bar{u}$, $\bar{v}$ such that $(\bar{u},\bar{x},\bar{y},\bar{v})$ satisfies the Kuhn-Tucker conditions

$$\begin{aligned} Ax + y &= b \\ -A^T u + Cx + v &= -c \\ u,x,y,v &\geq 0 \end{aligned}$$

(where T denotes transposition) and

$$uy + vx = 0 .$$

Further, from the above equations we have

$$\begin{aligned} c\bar{x} + \frac{1}{2}\bar{x}C\bar{x} &= c\bar{x} + \frac{1}{2}(-c\bar{x} + \bar{u}A\bar{x} - \bar{v}\bar{x}) \\ &= \frac{1}{2}(c\bar{x} + \bar{u}b - \bar{u}\bar{y} - \bar{v}\bar{x}) \\ &= \frac{1}{2}(c\bar{x} + \bar{u}b) . \end{aligned}$$

Therefore, denoting

$$\begin{pmatrix} 0 & A \\ -A^T & C \end{pmatrix} = M , \quad \begin{pmatrix} b \\ -c \end{pmatrix} = q , \quad \begin{pmatrix} u \\ x \end{pmatrix} = z , \quad \begin{pmatrix} y \\ v \end{pmatrix} = w ,$$

$$\frac{1}{2}\begin{pmatrix} b \\ c \end{pmatrix} = p ,$$

(P) can be restated as the linear program

$$\max\{pz \mid (z,w) \in Z \text{ for some } w\},$$

where

$$Z = \{(z,w) \varepsilon \mathbf{R}^n \times \mathbf{R}^n | Mz + w = q,\ z \geq 0,\ w \geq 0\},$$

with the added complementarity condition

$$zw = 0 \ .$$

More generally, the problem of finding a complementary (i.e., such that $zw = 0$) solution to a system of the form Z, usually called the linear compementarity problem, has received a lot of attention in the literature (see, for instance [20], [12]) and is known as a way of formulating not only quadratic programs, but also bimatrix games and other problems.

Theorem 6. Given a basic feasible noncomplementary solution to the system Z, let

$$z_i = a_{io} + \sum_{j \varepsilon J} a_{ij}(-t_j)$$

$$w_i = b_{io} + \sum_{j \varepsilon J} b_{ij}(-t_j)$$

be the expression of the i-th complementary pair of variables in terms of the nonbasic variables, and let $N = \{1,\ldots,n\}$, $N_* = \{i \varepsilon N | a_{io} > 0,\ b_{io} > 0\}$. Then every feasible complementary solution to Z satisfies the inequality

$$(31) \quad \sum_{j \varepsilon J} \alpha_j t_j \geq 1$$

for each of the following two definitions of its coefficients:

$$(32) \quad \alpha_j = \max \left\{ \frac{a_{ij}}{a_{io}}, \frac{b_{ij}}{b_{io}} \right\}, \text{ for any } i \varepsilon N_* \text{ (identical for all } j \varepsilon J)$$

and

$$(33)\quad \alpha_j = \max\left\{\max_{i\varepsilon Q} \frac{\alpha_{ij}}{\alpha_{io}} \, , \, \frac{\sum_{i\varepsilon Q} \bar{\alpha}_{ij}}{\sum_{i\varepsilon Q} \bar{\alpha}_{io}}\right\}, \text{ for any } \quad Q \subset N_*$$

(identical for all $j\varepsilon J$),

where for each $i\varepsilon Q$, $j\varepsilon J\cup\{0\}$, either $\alpha_{ij} = a_{ij}$ and $\bar{\alpha}_{ij} = b_{ij}$, or $\alpha_{ij} = b_{ij}$ and $\bar{\alpha}_{ij} = a_{ij}$.

Proof. The cut defined by (32) follows from applying Corollary 1.1 to any of the disjunctions $z_i \le 0 \vee w_i \le 0$, $i\varepsilon M$, stated in terms of the nonbasic variables. The cut defined by (33) can be obtained, on the other hand, from a disjunction of the form

$$(34)\quad u_{i_1} \le 0 \vee \ldots \vee u_{i_q} \le 0 \vee \{v_i \le 0, \forall\, i\varepsilon Q\}$$

where $\{i_1,\ldots,i_q\} = Q$ is any subset of N_* , and for each $i\varepsilon Q$, either $u_i = z_i$ and $v_i = w_i$, or $u_i = w_i$ and $v_i = z_i$. Restating (34) in terms of the nonbasic variables and applying Theorem 1, with the multiplier for the i-th inequality set equal to $\frac{1}{\alpha_{io}}$, $i = i_1,\ldots,i_q$, and all multipliers for the last system of inequalities set equal to $\left(\sum_{i\varepsilon Q} \bar{\alpha}_{io}\right)^{-1}$, produces (33).

Q.E.D.

The cuts of Theorem 6, while considerably cheaper to compute than those proposed in [3], do not use information from other problem constraints than the ones used to generate the cuts. However, a different choice of the multipliers can take into account the other constraints too.

References

[1] Balas, E. (1971) "Intersection Cuts - a New Type of Cutting Planes for Integer Programming," MSRR #187, Carnegie-Mellon University, October 1969. Published in Operations Research, 19, pp. 19-39.

[2] Balas, E. (1972) "Integer Programming and Convex Analysis: Intersection Cuts from Outer Polars." Mathematical Programming, 2, pp. 330-382.

[3] Balas, E. (July 1972) "Nonconvex Quadratic Programming." MSRR #278, Carnegie-Mellon University. Forthcoming in the SIAM Journal on Applied Mathematics.

[4] Balas, E. (August 27-31, 1973) "On the Use of Intersection Cuts in Branch and Bound." Paper presented at the 8th International Symposium on Mathematical Programming, Stanford, California.

[5] Balas, E. (February 1974) "Intersection Cuts from Disjunctive Constraints." MSRR #330, Carnegie-Mellon University.

[6] Balas, E. (In preparation) "Branch and Bound with Intersection Cuts from Disjunctive Constraints."

[7] Balas, E. (July 1974) "Disjunctive Programming: Properties of the Convex Hull of Feasible Points." MSRR #348, Carnegie-Mellon University.

[8] Balas, E. and Burdet, C.-A. (Sept. 1972 - July 1973) "Maximizing a Convex Quadratic Function Subject to Linear Constraints." MSRR #299, Carnegie-Mellon University.

[9] Balas, E. and Jeroslow, R. "Canonical Cuts on the Unit Hypercube." MSRR #198, Carnegie-Mellon University, August-December 1969. Published in SIAM Journal on Applied Mathematics, 23, 1972, p. 61-69.

[10] Balas, E. and Zoltners, A. (April 1973) "Intersection Cuts from Outer Polars of Truncated Cubes." MSRR #324, Carnegie-Mellon University. Forthcoming in Naval Research Logistics Quarterly.

[11] Burdet, C.-A. (1973) "Enumerative Cuts: I." Operations Research, 21, pp. 61-89.

[12] Cottle, R. W. and Dantzig, G. B. (1968) "Complementary Pivot Theory of Mathematical Programming." Linear Algebra and Its Applications, 1, pp. 103-125.

[13] Glover, F. (1973) "Convexity Cuts for Multiple Choice Problems." Discrete Mathematics, 6, pp. 221-234.

[14] Glover, F. (May 1973) "Polyhedral Convexity Cuts and Negative Edge Extensions." MSRS 73-6, University of Colorado.

[15] Glover, F. (August 1973) "Polyhedral Annexation in Mixed Integer Programming." MSRS 73-9, University of Colorado.

[16] Glover, F., and Klingman, D. (1973) "The Generalized Lattice Point Problem." Operations Research, 21, pp. 141-156.

[16a] Glover, F., Klingman, D. and Stutz, J. (June 1972) "The Disjunctive Facet Problem: Formulation and Solution Techniques" MSRS No. 72-10, University of Colorado.

[17] Gomory, R. E. (1960) "An Algorithm for the Mixed Integer Problem." RM-2597, The RAND Corporation.

[18] Hadley, G. (1964) Nonlinear and Dynamic Programming. Addison-Wesley.

[19] Jeroslow, R. (February 1974) "The Principles of Cutting Plane Theory: Part I." Carnegie-Mellon University.

[20] Lemke, C. E. (1968) "On Complementary Pivot Theory." G. B. Dantzig and A. F. Veinott (editors), Mathematics of the Decision Sciences. AMS.

[21] Mangasarian, O. L. (1969) Nonlinear Programming. McGraw-Hill.

[22] Owen, G. (1973) "Cutting Planes for Programs with Disjunctive Constraints." Journal of Optimization Theory and Applications, 11, pp. 49-55.

[22a] Owen, G. (1965) "A Cutting Plane Approach to the Fixed Charge Problem," unpublished research report for Mathematica, Princeton, N. J.

[23] Rockafellar, R. T. (1970) Convex Analysis Princeton University Press.

[24] Zwart, P. B. (January 1972) Intersection Cuts for Separable Programming." School of Engineering and Applied Science, Washington University, St. Louis, Mo.

A GENERALIZATION OF A THEOREM OF CHVÁTAL AND GOMORY*

by

R. G. Jeroslow

ABSTRACT

We generalize the result of Chvátal and Gomory, by providing two simple operations which, iteratively applied, yield all facets to the convex span of all feasible points to a linear problem plus an additional nonlinear constraint, when the linear constraints define a polytope. In integer programming, this additional nonlinear constraint is that the solution be integral. Our extension permits restrictions of the solution to more general sets than a bounded set of integer points. Understood algorithmically, it provides a finitely convergent cutting-plane algorithm for that class of nonlinear programming problems which, in addition to linear constraints, require that the solution lie in some finite set (e.g., a set of vertices).

1. Introduction

At the present time, only one non-trivial result is published concerning the facets of the

---

*This report was prepared as part of the activities of the Management Sciences Research Group, Carnegie-Mellon University, under Contract N00014-67-A-0314-0007 NR 047-048 with the U. S. Office of Naval Research. Reproduction in whole or in part is permitted for any purpose of the U. S. Government.

integer hull of a bounded polyhedron, i.e., concerning the convex span of the integer points in a polytope. This result is due to Gomory and was rediscovered independently by Chvátal.

Chvátal stated the result clearly in the form of a theorem [11]. This theorem is not stated explicitly in Gomory's paper [14], but it follows from Gomory's discussion of the nature of the cuts used in his algorithmic procedure (see particularly page 276 of [14] and the illustration on that page).

We now give Chvátal's statement of the result [11].

Consider a polytope of the form

$$(1) \qquad \begin{aligned} Ax &\le b \\ x &\ge 0 \,. \end{aligned}$$

Let IH be the integer hull of (1). We allow two operations on inequalities:

(i) The formation of non-negative linear combinations of given inequalities.

(ii) Given an inequality

$$c_1x_1 + c_2x_2 + \cdots + c_nx_n \le d$$

with all c_j integral, $j = 1,\ldots,n$, we may derive the same inequality with d replaced by $[d]$, where $[d]$ denotes the largest integer which is $\le d$.

Then Chvátal concludes that finitely many applications of (i) and (ii) can derive any supporting hyperplane to IH. Hence all the finitely many facets of IH can be derived in finitely many applications of (i) and (ii).

Gomory describes in [14] an algorithm which uses cutting-planes, and which finds the optimum to an integer program in finitely many steps. He gives a brief discussion which in effect shows that his cutting-planes involve only the operations (i) and (ii). Clearly, by starting with a linear criterion function and optimizing, the algorithm must terminate, when the integer program is consistent, with a supporting hyperplane to IH. As one varies the choice of criterion function, one can obtain all facets when IH is fully dimensional. Gomory [15] later extended his result to the mixed-integer case, when the criterion function is integer-constrained.

From an algorithmic viewpoint, this mathematical result gives no clue as to how one should pick the co-efficients of variables in order to begin with a hyperplane parallel to a facet which, following optimization, will intersect the facet. But practically, one is not interested in facets except as they lead to the solution of programming problems, and that is already obtained by the procedure for realizing supporting hyperplanes, which does constitute an algorithm.

In this paper, we shall generalize the theorem of Chvátal and Gomory, by specifying two operations for obtaining all valid linear inequalities satisfied by the convex hull of a set of points S which may have a definition far more complex than that of the integer vectors in a bounded region of space. We require only that S is finite.

The main tool which will assist us in our investigation will be asymptotic linear programming [16], a tool which we developed, following the introduction of the use of the Hilbert field in mathematical programming by A. Charnes (see pp. 750-751 of and 756-757 of [10]).

We shall assume that the reader is familiar with [16]. During our discussion we shall make sufficient remarks so that our argument is, at the least, quite "reasonable" for those who have not read [16].

Section 1: The Main Result

We have seen, in the introduction, that the programming language of optimizing linear forms subject to polytope constraints, or finding all linear inequalities satisfied by the convex span of certain points which satisfy the constraints, are alternate ways of discussing the same phenomena. We shall therefore feel free to move back and forth between the two formulations.

In programming language, we consider the problem:

$$(P) \qquad \begin{array}{ll} \text{maximize} & x_0 = cx + dy \\ \text{subject to} & Ax + Ey \leq b \\ & x \geq 0 \\ & (cx + dy, x) \in S , \end{array}$$

where S is a finite set, A and E are matrices, $x = (x_1,\ldots,x_p)$ is a vector, as is $y = (y_1,\ldots,y_q)$, c, d, and b are vectors of appropriate dimensions.

We single out the non-negativity conditions on x ; if there are also such conditions on y , these are assumed to be absorbed in the other inequality constraints. The non-negativity conditions on x play a technical role in the argument to follow. Also, it proves convenient to designate the criterion value $cx + dy$ as x_0 , although we feel free to discuss vectors $(x_0, x_1, \ldots, x_p)$ even if there does not exist a vector y such that $Ax + Ey \leq b$ and $x_0 = cx + dy$. Let Q be the polyhedron $Q = \{(x,y) | Ax + Ey \leq b, x \geq 0\}$.

We consider two operations on linear inequalities:

a) Taking non-negative linear combinations of given inequalities.

b) Given an inequality

$$ux + vy \leq d$$

such that
$d^* \geq \sup\{ux + vy | ux + vy \leq d, (cx + dy, x) \in S\}$
we may derive the same inequality with d replaced by d^* .

The following is our main result.

Theorem:[1] We assume that S is finite.

Suppose that all x_j , $j = 1, \ldots, p$, are bounded above in Q . Then the use of linear programming,

---

[1] This result was announced, for Q a polytope, at the April 15 session of the Nonlinear Programming Symposium sponsored by SIGMAP. An earlier version appears in [17, Section 5]. Other results announced at that session will appear in [18].

plus cuts derived via (a) and (b), is sufficient to solve the programming problem (P), if $cx + dy$ is bounded above in Q .

In alternate formulation, any inequality satisfied by the convex hull of the points of $Q \cap S$, can be obtained by finitely many applications of (a) and (b) , provided that the hyperplane of the given inequality can be translated so that Q lies entirely to one side, and provided that (P) is consistent.

Proof: Let B be so large a number that, whenever $(x_0,x) \varepsilon S$, we have $x_j - B < 0$ for $j = 1,\ldots,p$. B exists since S is finite.

We begin by constructing a set T , which is to consist of the union of S and all vectors of the form $(x_0,\ldots,x_{j-1},-B,x_{j+1},\ldots,x_p)$ such that $(x_0,\ldots,x_{j-1},x_j,x_{j+1},\ldots x_p) \in S$, for any $j=1,\ldots,p$. Clearly T is finite, and the first-named vector of the last sentence is called the j-th translate of the second-named vector $(x_0,x) \in S$, $x = (x_1,\ldots,x_p)$. Note that any element of $T\setminus S$ cannot be in Q , since all such vectors have at least one negative component.

We replace the criterion function $cx + dy$ of (P) by the criterion

$$(2) \qquad cx + dy + \sum_{i=1}^{p} x_i/M^i$$

where M is the infinitely large quantity of the Hilbert field of [16].

Since the quantities $x_j \geq 0$, $j = 1,\ldots,p$ are bounded above in Q in the real field, they remain bounded above by a real number in any ordered field, such as the Hilbert field, extending the reals [19]. Therefore the quantities x_1/M, $x_2/M^2,\ldots$ will be infinitely small, and so the criterion (2) and $cx + dy$ are equivalent. This shows that when (P) is bounded, it will also be bounded with (2) as criterion function in place of $cx + dy$. Since only the criterion function (2) contains infinite quantities, the simplex algorithm will yield only solutions (x,y) in the reals.

The infinitesimal quantities will exert a very specific control over which of the points with a given value of $cx + dy$ are visited by the (dual) linear programming algorithm.

The reader familiar with [16] may wish to verify, as an historical matter, that the use of the criterion function (2) is equivalent to the lexicographic dual method utilized in [14].

For an integer r in the range $0 \le r \le p$, we shall use

$$(3) \qquad (x_0,x_1,\ldots,x_r) < \text{lex } (\tilde{x}_0,\tilde{x}_1,\ldots,\tilde{x}_r)$$

to abbreviate the fact that there exists an integer s in the range $0 \le s \le r$ such that: 1) $x_i = \tilde{x}_i$ for $i = 1,\ldots,s-1$; and 2) $x_s < \tilde{x}_s$. In the case that $s = 0$, clause 1) is vacuous. We say that s is the <u>key index</u> of (3), and use $\le$ in place of $<$ in (3) to abbreviate that either $<$ or $=$ holds. The ordering of (3) is called "lexicographic," and such orderings occur in both the known proofs of the theorem of Chvátal and Gomory.

We leave it for the reader to check that (3) is equivalent to the algebraic condition

$$(3)' \qquad x_0 + \sum_{i=1}^{r} x_i/M^i < \tilde{x}_0 + \sum_{i=1}^{r} \tilde{x}_i/M^i$$

in the Hilbert field, when all x's are real numbers.

For any vector $(x_0,x) \in R^{P+1}$, $x = (x_1,\dots,x_p)$, we define the set

$$L(x_0,x). = \{(x_0',x') \in T \mid (x_0',x') \le (\text{lex})\ (x_0,x)\}.$$

If $L(x_0,x) \neq \emptyset$, it has a lexicographically largest element that we will call the <u>round-off</u> of (x_0,x). If $L(x_0,x) = \emptyset$, we say that the round-off "does not exist."

Now suppose that (P) is solved as a linear program, yielding the solution x^0,y^0. If $(cx^0 + dy^0, x^0) \in S$, we have found the optimum to (P). We assume in what follows that this is not the case, and show how to add cutting-planes via (a) and (b) which we will show yield finite convergence.

Put $x_0{}^0 = cx^0 + dy^0$. If the round-off of $(x^0{}_0,x^0{}_1,\dots,x^0{}_p)$ does not exist, (P) is inconsistent, and we terminate computation without the need of cutting planes. Otherwise, let $(\bar{x}_0,\bar{x}_1,\dots,\bar{x}_p)$ be the round-off of $(x^0{}_0,x^0{}_1,\dots,x^0{}_p)$. Since $(x_0{}^0,x^0) \notin S$, we have

$(\bar{x}_0, \bar{x}_1, \ldots, \bar{x}_p) <$ (lex) $(x^0{}_0, x^0{}_1, \ldots, x^0{}_p)$ if we assume that $(\bar{x}_0, \bar{x}_1, \ldots, \bar{x}_p) \in S$. Otherwise, $(\bar{x}_0, \bar{x}_1, \ldots, \bar{x}_p) \in T \backslash S$, so that $(\bar{x}_0, \bar{x}_1, \ldots, \bar{x}_p) \notin Q$ and $(\bar{x}_0, \bar{x}_1, \ldots, \bar{x}_p) \neq (x_0{}^0, x_1{}^0, \ldots, x^0{}_p)$; once again $(\bar{x}_0, \bar{x}_1, \ldots, \bar{x}_p) <$ (lex) $(x^0{}_0, x^0{}_1, \ldots, x^0{}_p)$. Let s be the key index of this $<$ (lex) relation.

Next we note that the inequality

$$cx + dy + \sum_{i=1}^{s} x_i/M^i \leq \bar{x}_0 + \sum_{i=1}^{s} \bar{x}_i/M^i \tag{4}$$

is satisfied by all feasible solutions, $(cx + dy, x) \in S$ (the summation (4) is empty, with value zero, in the case $s = 0$). This follows because, for such feasible solutions, $(cx + dy, x) \in L(x_0{}^0, x^0)$ when (2) is used as criterion function. Compare (4) with the inequality

$$cx + dy + \sum_{i=1}^{s} x_i/M^i \leq x_0{}^0 + \sum_{i=1}^{s} x_i{}^0/M^i \tag{5}$$

which is satisfied by all solution to the constraints $Ax + Ey \leq b$, $x \geq 0$, since the asymptotic linear programming algorithm has been used to obtain $(cx^0 + dy^0, x^0)$.

The polyhedron $Q = \{(x,y) \mid Ax + Ey \leq b,\ x \geq 0\}$ can be written as the sum $Q = H(A)+C(B)$ of the convex hull $H(A)$ of finitely many points A plus the convex span $C(B)$ of finitely many rays B, by the Finite Basis Theorem (see [19]). Every one of the points $a \in A$ satisfies (5), and everyone of the rays $b \in B$ must satisfy the homogeneous relation

$$(6) \qquad cx + dy + \sum_{i=1}^{s} x_i/M^i \le 0 \, .$$

Since $A \cup B$ is finite, for a suitably large real scalar θ, all elements of A satisfy

$$(5)' \qquad cx + dy + \sum_{i=1}^{s} x_i/\theta^i \le x_0{}^0 + \sum_{i=1}^{s} x^0{}_i/\theta^i$$

and all elements of B satisfy

$$(6)' \qquad cx + dy + \sum_{i=1}^{s} x_i/\theta^i \le 0 \, .$$

Hence, all elements of Q satisfy (5)' .

Therefore, by the corollary to the Kuhn-Fourier Theorem [17], the inequality (5)' derives from taking non-negative linear combinations of the constraints defining Q , plus possibly a weakening. However, since $(cx^0 + dy^0, x^0)$ satisfies (5)' with equality, there cannot occur a weakening; i.e., only the operation (a) is used.

There are only finitely many vectors $(x_0,\ldots,x_p) \in S$ such that $(x_0,\ldots,x_p) \le (\text{lex}) \; (x_0{}^0,\ldots,x_p{}^0)$. For each of these vectors we have $(x_0,\ldots,x_p) \le (\text{lex}) \; (\bar{x}_0,\ldots,\bar{x}_p)$ since $(\bar{x}_0,\ldots,\bar{x}_p)$ is the round-off of $(x_0{}^0,\ldots,x_p{}^0)$ and $S \subseteq T$. Hence, for each of these vectors, the inequality

$$(7) \qquad x_0 + \sum_{i=1}^{s} x_i/M^i \le \bar{x}_0 + \sum_{i=1}^{s} \bar{x}_i/M^i$$

holds. Hence, for each of these vectors, (7) holds with sufficiently large θ substituted through for M.

Next, consider any of the finitely many vectors $(x_0,\ldots,x_p) \in S$ such that $(x_0,\ldots,x_p) \geq (\text{lex})\ (x_0^0,\ldots,x_p^0)$. Let i be the key index of this relation. Then if $i \geq 1$ the i-th translate of $(x_0,\ldots,x_p)$ is $<$ (lex) $(x_0^0,\ldots,x_p^0)$ because its i-th component is negative, and hence the round-off $(\bar{x}_1,\ldots,\bar{x}_p)$ is $\geq$ (lex) this translate. Since this translate agrees with $(x^0_0,\ldots,x_p^0)$ in components zero to (i-1), so does the round-off, and hence $s \geq i$. If $i = 0$ $s \geq i$ is immediate. Therefore, $(x_0,\ldots,x_s) > (\text{lex})\ (x_0^0,\ldots,x^0_s)$, so that $>$ holds in (5)' in place of $\leq$ for all sufficiently large θ.

We conclude that the supremum d* of the linear form

$$x_0 + \sum_{i=1}^{s} x_i/\theta^i , \tag{8}$$

subject to $(x_0,x) \in S$ and the condition that (8) does not exceed the right-hand-side of (5)', is the right-hand-side of (7) with θ substituted for M. I. e., with one application of the operation (b), we validate the cutting-plane.

$$cx + dy + \sum_{i=1}^{s} x_i/\theta^i \leq \bar{x}_0 + \sum_{i=1}^{s} \bar{x}_i/\theta^i \tag{9}$$

for sufficiently large θ.

We now prove that only finitely many cuts need be added before either we find the optimum to (P) or discover that (P) is inconsistent. The proof is by contradiction, assuming that infinitely many cuts are added.

Since round-offs occur in the cuts in non-increasing lexicographic order, we cannot change round-offs infinitely often, or we would have a lexicographically decreasing sequence in the finite set T, which is impossible. Hence, after finitely many cuts have been added, the same vector $(\bar{x}_0, \ldots, \bar{x}_p) \in T$ recurs as round-off; in what follows, we hold this round-off fixed.

Only s can change in (9). Suppose some specific value s is used in (9). We shall now show that at least $(s+1)$ must be used in the next cut and, s being arbitrary, we have a contradiction after p more cuts, because s is bounded by p.

If s is used, then $x_i^0 = \bar{x}_i$, for $i = 1, \ldots, s-1$. Therefore, after the cut (9) is added, the new linear programming optimum (x, y) must have $x_i = x_i^0$, $i = 1, \ldots, s-1$, for if any of the first $(s-1)$ co-ordinates change, the least-indexed of these must decrease, causing a decrease in round-off, which is contrary to our assumption. Therefore, the cut (9) is equivalent to $x_s/\theta^s \le \bar{x}_s/\theta^s$, i.e., to $x_s \le \bar{x}_s$. Since $x_s < \bar{x}_s$ would cause a decrease in round-off, $x_s = \bar{x}_s$. As promised, the next key index must be $\ge (s+1)$. Q.E.D.

Remark 1: The algorithmic implementation of the procedure discussed in the proof of the theorem is as follows.

First, one solves (P) as a linear program, relaxing the nonlinear constraint $(cx + dy, x) \in S$. If the solution satisfies this nonlinear constraint, we are done. Otherwise, we compute the round-off $(\bar{x}_0, \bar{x}_1, \ldots, \bar{x}_p)$ of the linear programming solution $(x^0_0, x^0_1, \ldots, x^0_p)$ and add the cut (9) to the linear constraints of (P). The process is then repeated; the algorithm will terminate in finitely many steps.

Evidently, the algorithmic implementation will require an efficient method of computing round-offs, and this in turn depends upon the precise set S. For a set S without any structure, the round-off is, of course, purely theoretical, while it is trivial to compute for S a bounded set of integer points.

The choice of θ in (9) can be determined algorithmically as follows. One takes θ very large and determines if the linear program would lead to any more pivoting if the criterion function were

$$cx + dy + \sum_{i=1}^{s} x_i/\theta^i$$

If no pivoting away from $(x^0_0, x^0_1, \ldots, x^0_p)$ results, then θ is sufficiently large to insure (5)' and (6)'. Otherwise, one must increase θ , say, by doubling. In finitely many increases, we will have found a sufficiently large value of θ .

The cuts (9) of our theorem are, in a certain sense, the weakest possible cuts to solve the problem (P) for general S . When cuts which take advantage of the special structure of S are available, they are to be preferred.

Remark 2: The content of the Theorem is that the potentially quite difficult problem (P) can be replaced by finitely many simpler problems of the form discussed in (b). Our method of proof has been to give an "isomorphism" behind a lexicographic enumerative scheme and a cutting-plane scheme.

Remark 3: In order to obtain the theorem of Chvátal and Gomory for the case that the continuous variable y does not occur, we chose S to be all integer points in a large bounded region of space. The vector c is assumed a vector of integers for this theorem, and we chose θ to be a very large integer. Then rewriting (9) as

$$(11)' \qquad \theta^{s} cx + \sum_{i=1}^{s} \theta^{s-i} x_i \leq \theta^{s} \bar{x}_0 + \sum_{i=1}^{s} \theta^{s-i} \bar{x}_i \ ,$$

the definition of the round-off, plus the supposition that $(cx^0, x^0) \notin S$, shows that $\bar{x}_i{}^0$ is integral for $i = 1,\ldots,s-1$, while $x_s{}^0$ is fractional and $[x_s{}^0] = \bar{x}_s$. Hence, our operation (b) in this case amounts to Chvátal's operation (ii).

Remark 4: The proof of our theorem, when specialized to the integer programming context, reveals that the use of the lexicographic dual simplex method amounts to a "kind of" lexicographic search over the flat for a value cx^0 = constant. A full lexicographic search is not actually performed, since often $s < p$. A better comparison would be to an enumerative integer programming algorithm, in which the order of fixing variables has been chosen once and for all to be $x_1, \ldots, x_p$, and in which search nodes are only "temporarily fathomed" when they fail to give a certain desired criterion value, but may later be revisited for a different possible value of the criterion function.

This comparison would tend to explain the unusual performance of Gomory's algorithm [14] which has been observed in practice [21]. Specifically, when the algorithm of [14] obtains the optimum, it does so with great speed, or else it never appears to converge.

Gomory's original algorithm [14] would tend to find the optimum quickly if the optimal linear programming vertex were fairly "pointed" and the integer programming optimum were nearby. Otherwise, as the criterion plane cx = constant lowers and is passed through larger and larger cross-sections of the polyhedron Q, Gomory's algorithm would tend to enter a large lexicographic search. Once this happens, it is bound to be inferior to enumerative algorithms with better heuristic rules for choosing fixed variables than the rather blind lexicographic rule, even aside from the issue of visiting near the same integer point for many values of cx = constant.

This admittedly heuristic reasoning of ours co-incides with Woolsey's experience. He reports that with hindsight the least-cost way of implementing the detailed sensitivity analysis of [21] would not have been either to use Gomory's algorithm [14] exclusively, since it could not solve certain problems, nor an early version of Balas' additive algorithm [2] exclusively. On the problems solved by the cutting-plane algorithm, which were the bulk of the problems of [21], the algorithm of [2] was much slower.

Woolsey reported that the optimal strategy would have been to run all problems first with the cutting-plane algorithm and with a very short time limit, and then switch to the enumerative algorithm when the time limit was exceeded. His experimentation costs would have been reduced by a factor of about six under the optimal strategy.

It is Balas' opinion that researchers could profitably explore hybrid integer programming algorithms, such as enumerative algorithms assisted by cutting-planes [5], [20], and primarily cutting-plane algorithms in which some degree of enumeration occurs to validate stronger cuts [12].

<u>Remark 5</u>: We have just received (April 5, 1974) a paper of F. Glover [13] containing a result on cutting-planes which he mentioned to us in September, 1973 (private correspondence). A study of his proof reveals another characterization of the facets of the mixed-integer programming problem, which is also obtained by essentially an "isomorphism" with an enumerative scheme.

We shall discuss these matters in our paper [18], providing a theoretical generalization to certain unbounded sets S , but more importantly, giving methods which allow one to trade-off on the "degree of enumeration" versus the theoretical power of the cuts obtained. Other methods were given in [9] and [12]. The paper [4] gives several new and easily-computed cuts from the recent and very different new developments in cutting-plane theory.

Section 2: Acknowledgements

I have benefitted from the comments of Dr. Alan Hoffman of IBM, who pointed out to me that what later became known as the operations (i) and (ii) of Chvátal were sufficient to define the integer hull, and that this fact was the mathematical assertion corresponding to the algorithm described in [14].

I wish to particularly thank Professor Vasek Chvátal of Stanford University for many detailed comments on an earlier version [17] of the main result.

References

[1] Balas, E. (December 5-7, 1963) "Linear Programming with Zero-One Variables," (in Roumanian), Proceedings of the Third Scientific Session on Statistics, Bucharest, (published in English as 2.).

[2] Balas, E. (1965) "An Additive Algorithm for Solving Linear Programs with Zero-One Variables," Operations Research 13, 1965, pp. 517-546.

[3] Balas, E. (August 27-31, 1973) "On the Use of Intersection Cuts in Branch and Bound" a paper presented at the Eighth International Symposium on Mathematical Programming, Stanford, California. Parts of this talk have already appeared in 4. and other results from the talk will shortly appear in 5.

[4] Balas, E. (First draft, August 1973, Revised and expanded, February 1974) "Intersection Cuts from Disjunctive Constraints," Management Science Research Report No. 330, Carnegie-Mellon University, GSIA.

[5] Balas, E. (In preparation) "Branch and Bound with Intersection Cuts from Disjunctive Constraints."

[6] Bowman, V. J. and Nemhauser, G. L. (1971) "Deep Cuts in Integer Programming," Opsearch 8, pp. 89-111.

[7] Bowman, V. J. and Nemhauser, G. L. (1970) "A Finiteness Proof for Modified Dantzig Cuts in Integer Programming," Naval Research Logistics Quarterly, 17, pp. 309-313.

[8] Burdet, C.-A. (1972) "Enumerative Inequalities in Integer Programming," Mathematical Programming 2, pp. 32-64.

[9] Burdet, C.-A. (1973) "Enumerative Cuts: I," Operations Research 21, pp. 61-89.

[10] Charnes, A. and Cooper, W. W. (1961) Management Models and Industrial Applications of Linear Programming, Vols. 1 and 2, Wiley, New York.

[11] Chvátal, V. (1973) "Edmonds Polytopes and a Hierarchy of Combinatorial Problems," Discrete Mathematics 4, pp. 305-337.

[12] Glover, F. (1972) "Cut Search Methods in Integer Programming," Mathematical Programming 3, pp. 86-100.

[13] Glover, F. (March, 1974) "On Polyhedral Annexation and Generating the Facets of the Convex Hull of Feasible Solutions to Mixed Integer Programming Problems," Man. Sci. Report Series 74-2, University of Colorado.

[14] Gomory, R. E. (1958) "An Algorithm for Integer Solutions to Linear Programs," in Recent Advances in Mathematical Programming, Graves and Wolfe, eds., McGraw-Hill, 1963, pp. 269-302. Originally appeared as Princeton-IBM Math. Res. Project Tech. Report no. 1.

[15] Gomory, R. E. (1960) "An Algorithm for the Mixed Integer Program," RM-2597, RAND Corporation.

[16] Jeroslow, R. G. (1973) "Asymptotic Linear Programming," Operations Research 21, pp. 1128-1141.

[17] Jeroslow, R. G. (February 8, 1974) "The Principles of Cutting-Plane Theory: Part I", with Addendum. A xeroxed manuscript dated February 8, 1974, with Addendum dated February 13, 1974.

[18] Jeroslow, R. G. "Cutting-Planes for Relaxations of Integer Programs."

[19] Stoer, J. and Witzgall, C. (1970) Convexity and Optimization in Finite Dimensions: I Springer-Verlag, New York, 268+ pp.

[20] Tomlin, J. A. (1971) "An Improved Branch and Bound Method for Integer Programming," Operations Research 15. pp. 1070-1074.

[21] Woolsey, R. E. D. (August, 1969) "An Application of Integer Programming to Optimal Water Resource Allocation for the Delaware River Basin," Ph.D. dissertation, the University of Texas at Austin.

ZERO-ONE ZERO-ONE PROGRAMMING AND ENUMERATION

by

Monique Guignard[1]

Kurt Spielberg[2]

1. Introduction

It is possible to devise a special version of the Gomory all-integer integer algorithm so that: the introduced cuts are especially tight, the matrix of added cuts can be carried in logical form (has entries -1, +1, 0) and the current solution is on a vertex of the hypercube. The algorithm can always start and is likely to fail after a number of iterations. The circumstances under which failure occurs can be characterized to some extent, and one can use heuristic and linear programming devices to delay failure.

Aside from raising questions of intrinsic interest, the algorithm may have some use in branch and bound and enumerative programming. When failure occurs, the existing tableau can be taken as starting point for resolution of a linear program, and the resultant overall procedure qualifies for providing bounds in a branch and bound search. In this paper, however, we explore an enumerative procedure in which the final tableau (after failure) is taken to define a set of auxiliary cuts to be used in the enumeration.

1) Dept. of Statistics and O.R., Univ. of Pennsylvania
2) Philadelphia Scientific Center, IBM

2. All-integer zero-one programming.

Consider the zero-one program:

$$\begin{aligned} &\min \ c^o \cdot y \\ &C^o \cdot y \le b^o \\ &y(j) \text{ in } \{0,1\},\ j = 1,2,\ldots,n\ , \end{aligned} \tag{2.1}$$

in which C^o is a matrix of m rows and n columns, and b^o and c^o are correspondingly dimensioned vectors. The coefficients need not necessarily be integral, but computer tests of the proposed algorithms have been performed on integer data only.

Let (2.1) be rewritten in the form

$$\begin{aligned} \min z &= 0 + (-c^o)\cdot(-y) \\ s &= b^o + C^o\cdot(-y) \end{aligned} \tag{2.2}$$

$$\begin{aligned} \text{and } y &= 0 + (-1)\cdot(-y) \\ \bar{y} &= e + I\cdot(-y)\ , \end{aligned} \tag{2.3}$$

as suggested by R. E. Gomory in <1> (except for the last, redundant, constraint set).

I is an (n,n) identity matrix and e a vector of n entries 1 . $\bar{y}$ is the vector of complemented structural variables:

$$\bar{y}(j) = 1 - y(j) \tag{2.4}$$

(2.2) and (2.3) link the objective function z and the "dependent variables" $(s,y,\bar{y})$ with the "independent variables" $(-y)$ in terms of the matrix

$$A^o = \begin{pmatrix} 0 & -c^o \\ b^o & C^o \\ 0 & -I \\ e & I \end{pmatrix} \tag{2.5}$$

During an all-integer zero-one algorithm, at a general iteration, one has a matrix relation of the form (τ denoting transposition):

$$(z,s,y,\bar{y})^{\tau} = A \cdot (1,-t)^{\tau} \;, \tag{2.6}$$

with

$$A = \begin{pmatrix} v & -c \\ b & C \\ \gamma & Q \\ \bar{\gamma} & -Q \end{pmatrix} \tag{2.7}$$

The vector t has as elements the set of "current" independent variables, consisting of some composite of the structurals $y(j)$ and certain slack-variables $r(k)$ which are linear combinations of the original $y(j)$. A representation of the $r(k)$ in terms of the original $y(j)$ can be carried along easily:

$$r = \rho + R \cdot y \;. \tag{2.8}$$

Gomory's well-known all-integer integer algorithm maintains dual feasibility ($c(j) \geq 0$) and integrality of the "logical part" of A (the relations involving Q) . With suitable precautions in the selection of pivot rows and columns it has finite, albeit usually poor, convergence. Until optimality, the "current solution" (determined by $t=0$) has some negative components. As explained in <1>, one may derive from each corresponding row an implied pivot row

$$r(i) = -h(i) + (-g(i)) \cdot (-t) \tag{2.9}$$

with $h(i) > 0$ and $g(i,j) \geq 0$, determine the pivot column j to be the lexicographically smallest column with $g(i,j) > 0$, pivot $r(i)$ into the non-basis and discard the updated pivot row. ($r(i)$, $h(i)$ are scalars, $g(i)$ is an n-vector with entries $g(i,j)$). In the case of a zero-one program it is natural to take the parameter λ of Gomory, which determines the values of the pivot row coefficients, large enough to render $h(i)$ one and the $g(i,j)$ zero or one. The pivot step then becomes particularly simple, consisting in subtraction of the pivot column from all other columns with a minus one entry in the pivot row.

To obtain a stronger algorithm we suggest the use of the reduction procedures of <2>, as follows:

(a) Drop the objective function row and all feasible rows (those with nonnegative constant term).

(b) Drop all positive terms in the resulting matrix and multiply each column with the range (bound interval) of the corresponding variable (initially 1).

(c) Reduce the resulting matrix as in <2>, to obtain a set of minimal preferred inequalities.

These inequalities have the form (2.9) with all coefficients $(-h(i),-g(i,j))$ equal to 0 or -1, and it is clear that for any Gomory inequality as described above there exists a minimal preferred inequality which is either equal to it or better, in the sense of having fewer coefficients.

To have fewer coefficients in the pivot rows than usual is important in that it obviously delays the onset of the well-known degeneracy situations which impede the Gomory all-integer algorithm in practice.

A simple example may be useful. Let all $t(j)$ have range 1 and let an infeasible inequality of (2.7) be:

$$s = -9 + 3\ t1 + 5\ t2 - 4\ t3 + t4\ , \text{ or}$$

$$3\ t1 + 5\ t2 - 4\ t3 + t4 \geq 9\ .$$

Then the Gomory cut with λ large is:

$$t1 + t2 + t4 \geq 1$$

and a minimal preferred inequality is:

$$t2 \geq 1$$

3. <u>(Pseudo) zero-one zero-one representation</u>.

The representation of the dependent variables in terms of the independent variables will be said to be "zero-one zero-one" as long as Q has entries 1, -1, 0 and γ has values 0 and 1 only. A computer code could then keep the logical part of A in the form of two bit matrices (one for the non-negative and one for the nonpositive entries) of dimension (2n,1+n).

Let A' consist of the first (1+m+n) rows of A, after replacement of any row in the (γ,Q) portion of A with $\gamma(i) = 1$ by the corresponding row in the $(\bar{\gamma},-Q)$ submatrix of A. The last n rows of A' thus have constant terms 0. Hence one has a representation of the form

$$
\begin{aligned}
z &= v + (-c) \cdot (-t) \\
s &= b + \quad C \cdot (-t) \\
\tilde{y} &= 0 + \quad L \cdot (-t) \text{ , or}
\end{aligned}
\tag{3.1}
$$

$$
A' = \begin{pmatrix} v & -c \\ b & C \\ 0 & L \end{pmatrix}
\tag{3.2}
$$

A given row i of $(0,L)$ represents either $\tilde{y}(i) = y(i)$ or $\tilde{y}(i) = 1 - y(i)$ in terms of the current non-basic variables t .

It will be necessary to consider L partitioned (after rearrangement of columns) into:

$$
L = (P,Z,N) \text{ ,}
\tag{3.3}
$$

P having at least one $+1$ in each column and Z and N having only non-positive entries. The columns in Z are those with zero reduced costs (or with unacceptably small reduced costs). Within P , and also within (Z,N), let the columns be ordered in terms of increasing reduced costs.

As discussed in section two, a pivot row is generated from a row of (b,C) with $b(i) < 0$. It has a -1 in the constant column and 0 or -1 coefficients elsewhere. The pivot step alters A' by subtracting the pivot column from all other columns with corresponding -1 entries in the pivot row. The pivot row itself is auxiliary and is not retained.

The following properties are worth noting: (assume that $c>0$)

1. The initial tableau, corresponding to (2.2) and (2.3), is a zero-one zero-one representation,

with P void, and (Z,N) = -1. Moreover, any pivot step of the all-integer zero-one algorithm will lead to a second tableau which is a zero-one zero-one representation.

2. Given a zero-one zero-one representation at a general iteration, consider a pivot step based on a pivot row with non-zero entries in (Z,N) only. The pivot column and the columns to be modified have no positive entries in L , and the pivot step will lead to another zero-one zero-one representation.

3. A necessary and sufficient condition for a pivot step to yield another vertex of the hyper-cube (i.e., a relation with $\gamma(i) = 0$ or 1) is that the pivot column be not in P .

If a pivot row can be constructed with 0 entries outside of N , the pivot column will be in N , and the pivot step will not only lead to another zero-one zero-one representation, but will also result in an increase in the objective function (will not be degenerate).

It will be noted that necessary conditions for 2. can not be stated easily. While any deviation of the L matrix from the (1,-1,0) format will derive from a pivot step with a -1 pivot row coefficient in a column of P , some matrix entry changing to (+1)-(-1), such a change will only take place if there is a -1 pivot column coefficient in the row of the +1 entry about to change. Given a particular pivot row candidate, one can check for this case.

But that is not likely to help much in a general procedure for constructing "favorable" pivot rows, i.e. pivot rows which maintain zero-one zero-one representation.

4. (Pseudo) zero-one zero-one algorithms.

A zero-one zero-one algorithm would be characterized by a sequence of zero-one zero-one representations of the dependent variables. Geometrically one would move along edges of the polyhedron determined by (2.1) and the cuts of the all-integer algorithm from one (infeasible) vertex of the hypercube to the next. Attaining a feasible vertex would amount to termination of the algorithm.

The prefix (pseudo) is meant to make it clear from the outset that such a goal can hardly ever be attained. But one might hope to delay the failure of the zero-one zero-one procedure until a point at which the number v might have a usefully large value (say, a value larger than the objective function value of the relaxed linear program over (2.1)) by virtue of the following observation:

> All difficulties are related to the presence of a -1 in the pivot row in column j of the P matrix, and any such given -1 entry can be removed. One need merely replace the pivot row by the sum of the pivot row and one of the rows of L which have a $+1$ entry in column j. The resulting row has a zero in the column in question and retains the constant term of -1.

If necessary, reduction may then be used to give a new pivot row which is guaranteed not to have a -1 entry in the initially offending column j.

Of course this does not prevent the presence or creation of another negative pivot row entry in a column of P. What is needed is a systematic procedure for elimination of all such entries.

Such a procedure will involve the creation of a linear combination of given rows to be, eventually, added to a pivot row with undesirable entries of -1. If the linear combination can be made to have a negative constant term and positive coefficients everywhere else, or wherever needed, then the resulting final pivot row will have no undesirable entries.

We shall describe two procedures. One can not expect them to be fully successful. Typically, failure occurs when P (or some transformed version of P) contains a "cycle" such as:

```
+1      -1
-1  +1
    -1...+1
```

The zero-one zero-one algorithm will be similar to the all-integer zero-one algorithm given in section 2. What distinguishes it is an attempt at constructing a pivot row with no -1 entry in a column of P (or of P and Z).

Consider two "transformation" procedures with four typical steps:

1. Select a submatrix T of $D = \begin{pmatrix} C \\ L \end{pmatrix}$

   with rows indexed by I(T):

   either (i) the infeasible rows of C and all rows of L

   or (ii) all rows of D .

   and with columns indexed by J(T):

   either (i) all columns of P ,

   or (ii) all columns of (P,Z).

2. Order the columns of T , columns of P always preceding all other columns.

3. If possible, construct an (at least partial) non-negative linear combination of the rows of T , say u . T .

4. Using the result of 3 , add u . [d(I(T)), D(I(T))] to the pivot row so as to produce a new pivot row with no undesirable -1 coefficients $(d = \begin{pmatrix} b \\ 0 \end{pmatrix})$.

Procedure 1:

Use linear programming. Determine a non-negative multiplier u subject to:

$$u \cdot T \geq e \qquad (4.1)$$
$$u \cdot d(I(T)) \leq 0 \quad .$$

Procedure 2:

Use a heuristic procedure in which a sequence of matrices

$$T(0) = T, T(1)=M(1)T(0), T(2), \ldots, T(k)=M(k)T(k-1) \qquad (4.2)$$

is constructed by nonnegative linear transformations $M(j)$, $j < k$, such that $T(k)$ has all entries in columns 1 through k positive.

This can certainly be done for $k = 1$, since column 1 of T contains a plus 1 entry from P. The procedure can be continued until a column k in $T(k)$ ($k \leq$ no. of columns of T) contains no positive entry.

<u>Details and possible modifications.</u>

The above has been kept somewhat vague since it is not completely clear what constitutes a best approach.

First of all, it is not obvious in what order one should:

(i) transform

(ii) generate possible pivot rows by reduction (in the sense of <2>).

In our implementation we first executed the heuristic transformation procedure on $T = (P,Z)$ with all rows of L and the infeasible rows of C, followed by reduction of the matrix product of $M(k)$ by the $I(T)$ rows of (d,D). When the resulting pivot rows were all unsuitable, we made another attempt (rarely successful) at removing -1 entries in a given row $r(i)$ by addition of suitable rows in P.

We ordered $T(0)$ only, keeping the columns of P corresponding to nonbasic $r(i)$ leftmost, followed by the nonbasic $y(i)$ and then by the $r(i)$ and $y(i)$ of Z.

A possibly better procedure might be:

(i) Generate a number of minimal preferred inequalities, i.e. a number of possible pivot rows $r(i) = -h(i) + g(i).t$. For every such candidate:

(ii) Form T as above, but with the corresponding entries of $g(i)$ adjoined as an additional, say last, row.

(iii) At step j in the transformation leading to the sequence (4.2), reorder $T(j)$ according to the last row, keeping essentially the negative entries at the left.

The generation of the sequence could then possibly terminate successfully with $k < n$ when no negative entries remain in the last row of $T(k)$.

Note that the procedure 1 can also be improved by a weakening of the constraints. I.e., if (4.1) is infeasible, one can relax the problem by changing to 0 entries of 1 in e corresponding to columns j of T with no negative coefficient $(-g(i,j)) = -1$ in the pivot row. Finally, for each $r(i)$ in the modified procedure, with pivot column j not in P, one may check whether a pivot step to another zero-one zero-one representation is after all possible (see the remarks concerning the necessity of the conditions in section 3).

5. Applications, Enumeration.

In the following, we assume that the zero-one algorithm cannot be or is purposely not continued (terminates) at iteration p .

The utility of the algorithm depends almost entirely on the value of the objective function bound v at termination. If it is reasonably large compared with the objective function of the linear programming relaxation for (2.1), then a number of applications are possible and attractive.

Such applications would rest upon the fact that the solution $(z,y)^p$ at termination step p could be considered a solution of a different relaxed version of (2.1). As such, it could for example be used in branch and bound (BB) schemes.

At each node of the BB procedure one could take z^p as the objective function bound and could utilize the tableau for guidance in branching. The integrality and zero-one zero-one nature would be important (certainly on machines with large ratios of floating point to fixed point execution times).

Another possibility at termination is the dropping of rows in P which lead to difficulties. This has the effect of relaxing the problem further, with the compensating advantage of permitting the zero-one zero-one algorithm to proceed. For the 6 by 12 test problem we can attain the integer problem objective function value ($z = 13$) in such a fashion.

In this section we shall briefly concentrate on enumerative procedures using one termination tableau only. We use three sets of data:

1) The initial tableau: Y - representation.

2) The zero-one zero-one representation at termination: YR - representation. (the independent variables t are a known mixture of y(j) and r(k)).

3) The relation between the introduced vector r = (r(k)) of slack variables and the structural variables y(j).

$$\begin{array}{llll} \text{Y:} & \min c^{o} \cdot y & \text{YR:} & \min c \cdot t \\ & C^{o} \cdot y \le b^{o} & & C \cdot t \le b \\ & & & L \cdot t \le 0 \\ & & & t..\text{given vector of} \\ & & & \quad y(j) \text{ and } r(k) \end{array} \tag{5.1}$$

$$r = \rho + R \cdot y \tag{5.2}$$

To render the enumeration more effective, we also solve the original problem as a linear program followed by imposition of Gomory-Johnson cuts (see e.g. <3>,<4>) and retain the toprow of the final tableau. It is used throughout the enumeration in "ceiling tests", i.e. comparisons with an incumbent objective function bound z* for the purpose of fixing variables or shrinking bound sets. The use of this simple procedure is probably not sufficiently widespread. It provides an excellent and inexpensive vehicle for taking advantage of cutting plane techniques.

We implemented two types of enumeration:

(i) Enumeration over y (direct use of the Y-tableau, the YR - tableau being used as auxiliary tableau of cuts (similar to the cuts of Benders <5>)),

(ii) Enumeration over the YR - tableau, with the Y - tableau used as auxiliary constraints.

In both procedures the equations (5.2) were used to provide updated bounds on the r(k) and also, when indicated, to furnish additional cuts.

The enumerations made consistent use of reduction procedures proposed in <2> to fix variables and to shrink bounds whenever possible. For an alternate approach see <9>. Minimal preferred variables were then generated and utilized. It can be shown that this involves (for procedure (ii)) multiplying a column j of the YR - tableau, corresponding to a slack variable r(k), with the current bound interval for r(k) before reduction. (See Appendix 1).

A typical minimal preferred inequality might then be interpreted as a disjunctive logical constraint of the form: either ... lower the upper bound of t(j1), or ... raise the lower bound of t(j2), or .... .

A "branch" would be selected from such a disjunctive constraint. Priority would be given to possible branches with the property of "double contraction", i.e. to branches which would satisfy one disjunctive constraint and contradict another.

The enumerations were carried out so as to associate a forward branch always with a raising of a lower bound. Such a restriction is of course not necessary. (For details see the Appendix 2)

We feel that an enumeration scheme of the above type should be among the better possible ones, as long as the reduction procedure is not expensive (this would depend on efficiency of implementation, computer execution times and similar considerations). As pointed out elsewhere, e.g. in <2>, there are important classes of mixed integer problems, with (5.1) representing a set of Benders inequalities, for which the cost of reduction would be negligible compared with the other required data processing operations.

Finally, it should be remarked that "state enumeration procedures" (see <6>, <7>, <8>) would find natural application here. Our implementation, however, did not attempt to define states.

6. Experimentation and results.

We summarize a set of experiments on five test problems (described elsewhere, e.g. in <10>), in the TABLE.

The (10,20) and (10,28) problems are maximization problems treated as minimization problems with negative cost coefficients. However, our programs require dual feasibility, which is imposed by the introduction of the complementary variables $\bar{y}(j) = 1 - y(j)$. This explains the negative constant term in column 8 (also present in the initial tableau of the linear program).

The first five columns describe the problem and an initially performed linear program with a number of added Gomory-Johnson cuts (see column 5).

TABLE

(PSEUDO) ZERO-ONE ZERO-ONE PROGRAMMING PLUS ENUMERATION

1 PROBLEMS	2 OPTIMAL OBJ.F. OF INT. PROBLEM	3 OPTIMAL OBJ.F. OF LP	4 OPTIMAL OBJ.F. OF LP + CUTS	5 NO OF CUTS	6 Z* IMPOSED UPPER BOUND	7 NO OF ALL INTEGER PIVOTS z^P max	8 ALL INTEGER OBJ.F. AT START OF PSEUDO	9 ALL INTEGER OBJ.F. AT TERM. OF PSEUDO	10 Y-ENUM. NODES TO TERMINATION WITHOUT USE OF PSEUDO TABLEAU	11 Y-ENUM. NODES TO TERMINATION WITH USE OF PSEUDO TABLEAU
(6.12)	13	6.85	8.94	5	99	5	0	10	9	9
(10,20)	-6120	-6155.3	-6155.3	0	-6100*	10	-8655	-6590	75	11
(10,28)	-12400	-12462	-12433	4	-12300	12	-15495	-14095	43	27
(28,35)	550	521.05	542.6	2	575	22	0	425	43	21
(12,44)	73	56.7	62.9	3	77	6	0	33	439	299
(12,44)	73	56.7	69.8	7	77	2	0	24	85	45

For a description of such cuts the user may wish to consult <3> and references cited in <4>. Gomory-Johnson cuts are generated from all fractional rows associated with structural integer variables. In the generation process (due to Johnson), the slacks are treated as integer variables (of course only for pure integer programs).

As described before, the top row of the final tableau (column four) is retained for use by the final enumeration procedure. It allows the fixing of variables, or shrinking of variable bounds, both for non-basic integer and slack variables. We also extract a Benders inequality from the top row (i.e., an inequality over the nonbasic structural variables) and append it to the initial integer tableau to be used during the pseudo zero-one zero-one procedure. The Gomory-Johnson cuts therefore are of importance throughout the experimental runs.

Column 6 lists an upper bound z^*, imposed throughout all computations. One excepttion to this is marked by a *. For the (10,20) problem the tight upper bound of -6100 makes the problem too easy for the enumeration. It is therefore replaced for that purpose by -5500.

Column seven describes how many zero-one zero-one representation tableaux were generated by the algorithm, and column nine gives the value z^p_{max} (the last objective function value). Except for the first small problem, these values are disappointingly small compared to those in columns 3 and 4.

However, it should be mentioned that experiments without transformations 4.2 yielded results which were much worse. The transformation procedures do permit a substantial increase in the figures of columns 7 and 9.

To improve matters, one would have to implement some of the suggestions of chapter 4. We do not believe that one will be able to go very much further, but we do not really know. Among other things there is always the difficulty of erratic behavior of the best planned algorithms. The last two rows of the TABLE, e.g., show in column 7, that the use of stronger cuts (more Gomory-Johnson cuts ought to lead to a stronger Benders inequality) can influence the zero-one zero-one procedure unfavorably. One can always observe isolated instances of such behavior and must trust that the rational procedure is the good one in the large majority of cases.

The matrices R consist initially of (0,1) rows. But the algorithm eventually generates rows with large integer coefficients. The buildup in upper bound values for the r(i) is extraordinarily rapid. Large bounds are clearly a determining factor in the bad convergence of the dual all-integer algorithm. Convergence of that algorithm requires that feasible values of the t(j) be found, and these values will often be very large. From this point of view, the very fact of finite convergence is remarkable.

YR - enumeration will deteriorate with increasing bounds. Experimentation shows this behavior quite clearly. Even small values of p may already be too large. For some problems we found that the efficiency of the YR - algorithm versus p was roughly a bell-shaped function with an optimum below the maximal value we were able to obtain (column 7).

For Y - enumeration, on the other hand, it appeared that the most important factor was the attained objective function value in column 9.

Our main result is the comparison in the last two columns, the first dealing with Y - enumeration without auxiliary cuts, the second with Y - enumeration in which reduction procedures were applied to auxiliary cuts represented by the last zero-one zero-one representation tableau. As can be seen, the improvement in terms of nodes to be evaluated can be substantial.

7. Appendix:

1. Reduction procedures for bounded variable inequalities.

Consider the inequality:

$$\sum_{J} c(i,j).t(j) \leq b(i), \tag{A1}$$

with $$L(j) \leq t(j) \leq U(j). \tag{A2}$$

The transformation:

$$\tau(j) = (t(j)-L(j))/(U(j)-L(j)) \tag{A3}$$

leads to variables in the hypercube:

$$0 \leq \tau(j) \leq 1 \tag{A4}.$$

After the transformation A3, a typical inequality A1 takes the form:

$$\sum_{J} d(i,j).\tau(j) \leq \beta(i) \qquad \text{(A5)}$$

with $d(i,j) = c(i,j).(U(j)-L(j))$

and $\beta(i) = b(i) - \sum_{J} L(j) \, . \, c(i,j)$.

In the following we assume that these transformations have been carried out. A generalization of the reduction procedure of <2>, to be described more fully in <11>, can then be applied to the transformed data. It associates with each constraint (A5) a minimal preferred inequality of the form

$$\sum_{\pi(i)} \tilde{\tau}(j) > 0 \, . \qquad \text{(A6)}$$

$\pi(i)$ is a set of "minimal preferred variables" and $p(i) = |\pi(i)|$ is called the "preferred weight" of row i (the term "minimal" refers to p(i); what is minimized by the reduction algorithm is the number of terms in $\pi(i)$).

When all $\tau(j)$, $j \in \pi(i)$, are bivalent variables, then (A6) yields the normal interpretation: at least one of the variables $\tilde{\tau}(j)$ must be one if the original inequality is to be satisfied.

When the $\tau(j)$ are continuous variables, $0 \leq \tau(j) \leq 1$, $j \in \pi(i)$, then the interpretation of (A6) is the following:
not all $\tilde{\tau}(j)$, $j \in \pi(i)$, can be zero, therefore not all $t(j)$, $j \in \pi(i)$, can simultaneously be at their lower (upper) bounds. One of the bounds at least must be raised (lowered) by no less than one unit.

If $p(i) = 0$, the problem is infeasible.

If $p(i) = 1$, then $\tilde{\tau}(j)$, $j \in \pi(i)$, must be positive, and $L(j)$ must be raised by one unit if $\tilde{\tau}(j) = \tau(j)$, or $U(j)$ must be decreased by one unit if $\tilde{\tau}(j) = 1 - \tau(j)$.

In the general case $p = \min_i \{p(i)\} > 1$, one retains all minimal inequalities of size p. Suppose there are k such minimal preferred inequalities. They can be represented in a k by n matrix Q . Each row i of Q has p nonzero entries in the positions j of the associated preferred variables. A +1 entry identifies the case $\tilde{\tau}(j) = \bar{\tau}(j)$, a -1 the case $\tilde{\tau}(j) = \tau(j)$.

The matrix Q can be used to give direction to a search procedure. Examples of such use can be found in Appendix 2, also in references <2> and <12>.

2. The essentials of the enumerative procedure.

As explained before, the enumerative procedure was coded so as to permit enumeration either over the Y - tableau or over the YR - tableau. Let x correspondingly stand for either y or t . Appropriate comments will be made where the algorithm differs for y and t . Capitalized names refer to subroutines to be described further.

1. Set the $x(j)$ to the given lower bounds $L(j)$ for all j . Label all variables "free". Set the node count and the "level" of the search to 0 .

2. Node computation.

   2.1 Increase the node count.

   2.2 Compute current working bounds in WORKB.

   2.3 Ceiling test in CEILING.

   2.4 Reduction (fixing of variables, shrinking of bounds) in REDUCE.

   In 2.2 to 2.4 the following conditions may obtain:

   (i) The current node may be "fathomed". That means: the node (with its successor nodes) is established as infeasible, or

   the node is associated with an integer solution (by virtue of the chosen origin of the search and dual feasibility of the tableaux no successor node can have an improved integer solution), or

   the variables are all fixed (zero bound intervals).

   In such a case, go to 4 ("backup").

   (ii) Variables can be fixed or bound intervals can be shrunk. In such a case, go to 2.2.

3. Forward step ("Branch").

   3.1 Establish a constraint set $A \cdot x \leq b$, either from the Y - tableau or from the YR - tableau, depending on the chosen type of enumeration. As explained before, columns of bounded variables must be multiplied by the appropriate (working-) bound interval.

3.2 Compute a "minimal preferred inequality matrix Q" from the constraint set (by the methods of <2>; see also Appendix 1).

Let J be a set of indices for possible branching.

If Q is empty (which is only rarely the case), set J equal to the set of indices of the free variables (the variables not yet fixed). Go to 3.4.

Otherwise, row i of Q may be interpreted as follows:

Aside from zero entries it has exactly $p > 1$ entries of +1 or -1 . The smaller a value of p (the "preferred" weight), the more constrained is the problem.

$q(i,j) = +1$ (-1) means that constraint i can be satisfied (indeed must be satisfied for at least one j) by a lowering of the upper bound (raising of the lower bound) on $x(j)$.

3.3 If there are "double-contracting" variables $x(j)$, i.e. variables such that for j there exist i1 with $q(i1,j) = 1$ and i2 with $q(i2,j) = -1$, have J include all corresponding indices j . Go to 3.4.

If not, let I be the set of indices of rows of Q such that all nonzero entries are -1 .

If $T \neq 0$, set $J = \{j | q(i,j) = -1, i \in I\}$. Go to 3.4.

If I is void, drop all positive coefficients in A . Go to 3.2.

(The purpose of this step is to force a matrix Q with I not void above, so that the preferred variable inequalities suggest a "Forward Step" corresponding to a raising of a lower bound. Of course, the algorithm could be, and probably should be, generalized to allow forward steps which consist of the lowering of some upper bounds).

3.4 For all $j \in J$ compute the new right hand side (of the Y tableau or the YR - tableau). Choose the branch variable j^* so as to minimize overall infeasibility.

3.5 Record the forward step:

Add j^* to an index list, record $L(j^*)$, the lower bound of $x(j^*)$ before the forward step, in a bound list. Increase the level ℓ by one.

Go to 2.

4. Backward step.

Retrieve $j^*(\ell)$ and $L(j^*(\ell))$ from the index and bound lists. Fix the upper bound of $x(j^*(\ell))$ at $L(j^*(\ell))$.

Replace ℓ by $\ell - 1$. Terminate if $\ell < 0$.

Otherwise, go to 2 .

Subroutines:

WORKB:

1. After a forward step, go to 3 .

2. Reconstruct bounds from the current index and bound list.

3. When $x = t$, special "state" vectors for y are maintained by the program (giving for each fixed $y(j)$ its value and the level at which it was fixed).

   This information is used together with equation (2.8) to compute bounds on those $t(j)$ which are slacks $r(k)$ of the pseudo zero-one zero-one procedure.

4. Compute the right hand sides of both tableaux (Y and YR), as well as of the saved top row of the initial linear program (+cuts), by substituting the lower bounds of y and t , respectively.

   If one of the tableaux yields a feasible solution, compare it to the incumbent (currently best) solution, and record it if it is an improvement.

REDUCE:

Given a set of working bounds on y and x , there exist simple procedures (e.g. <9> or <2>) for deriving possible further restrictions on the bounds. Using these procedures, REDUCE alters bounds and fixes variables when possible. It must be emphasized that REDUCE uses both tableaux, as well as the relations 4.2, regardless of the type of enumeration.

CEILING

The top row of the initial linear program (+cuts) is of the form $z = z^o + a \cdot \sigma$, with $a \geq 0$, σ a vector of integer and slack variables, and z^o the objective function value given in column 4 of the TABLE. Given an upper bound z^* on the integer objective function and bounds on σ, the condition $z \leq z^*$ may permit the computation of improved bounds on the components of σ and y.

8. References

[1] Gomory, R. E. (1963) "All-Integer Integer Programming Algorithm," in Graves, Wolfe, "Recent Advances in Mathematical Programming," McGraw Hill.

[2] Spielberg, K. (1972) "Minimal Preferred Variable Reduction for Zero-One Programming", IBM Philadelphia Scientific Center Report 320 3013.

[3] Gomory, R. E., Johnson, E. L. (August 1972) "Some Continuous Functions Related to Corner Polyhedra", Mathematical Programming.

[4] Johnson, E. L., Spielberg, K. (December 1971) "Inequalities in Branch and Bound Programming," IBM Research Report RC 3649. Also in: "Optimisation Methods", Editors: Cottle, Krarup, English Univ. Press, London, 1974.

[5] Benders, J. F. (1962) "Partitioning Procedures for Solving Mixed-Variables Programming Problems," Numerische Mathematik.

[6] Guignard, M. M., Spielberg, K. (1968) "Search Techniques with adaptive Features for certain Integer and Mixed-Integer Programming Problems", Proc. IFIPS Congress, North-Holland Publ. Co.

[7] Guignard, M. M., Spielberg, K. (June 1973) "A Realization of the State Enumeration Procedure", IBM Philadelphia Scientific Center Report 320 3025.

[8] Guignard, M. M., Spielberg, K. (1972) "Mixed-Integer Algorithms for the (0,1) Knapsack Problem," IBM Journal of Research and Development.

[9] Zionts, S. (1972) "Generalized Implicit Enumeration Using Bounds on Variables for Solving Linear Programs with Zero-One Variables", Naval Research Logistics Quarterly.

[10] Lemke, C. E., Spielberg, K. (1967) "Direct Search Zero-One and Mixed Integer Programming", Operations Research.

[11] Guignard, M. M., Spielberg, K. (to appear) "Reduction Procedures for Bounded Variables Integer Programming".

[12] Spielberg, K. (June 1973) "A Minimal Inequality Branch- Bound Method", IBM Philadelphia Scientific Center Report 320 3024.

SUBJECT INDEX

A 5 B 6 C 7 D 8 E 9 F 0 G 1 H 2 I 3 J 4